Genetics—Principles and Perspectives: a series of texts

General Editors: Dr. K. R. Lewis, Professor Bernard John

Volumes already published

1 **The Differentiation of Cells** Norman Maclean
2 **The Genetics of Recombination** D. G. Catcheside, F.R.S.
3 **Chromosome Genetics** H. Rees, F.R.S. and R. N. Jones

The Genetics of Recombination

D. G. Catcheside, F.R.S.

Formerly Professor of Genetics, Australian National University, Canberra

University Park Press

First published 1977
by Edward Arnold (Publishers) Limited
25 Hill Street, London W1X 8LL

First published in the USA in 1977 by
University Park Press
233 East Redwood Street
Baltimore, Maryland 21202

Library of Congress Cataloging in Publication Data

Catcheside, D. G.
The genetics of recombination.

(Genetics, principles and perspectives; 2)
Bibliography: p.
Includes index.
1. Genetics recombination. I. Title. II. Series.
[DNLM: 1. Recombination, Genetic. QH443 C357g]
QH443.C37 1978 575.1 77-16201
ISBN 0-8391-1196-7

Printed in Great Britain

Preface

Recombination is genetic analysis, the fourth and distinctive mode of biological study, the others being morphological description, chemical and physical methods of analysis. Functionally, recombination is commonly concerned in ensuring the accuracy of chromosome segregation during meiosis. It is implicated in the molecular parasitism of temperate bacteriophages initiated by the insertion of their DNA into those of their hosts. It is a mechanism for generating and conserving genetic variability so that new combinations are subjected to the tests of natural selection. It is the hallmark of sexual reproduction.

The purpose of this book is to examine the process itself. To do so fully at the molecular level proves to be difficult with present knowledge. Nevertheless the shape of the solution to the problem seems reasonably clear. In spite of the diversity of the derivative functions, which are the products of evolutionary advantage, it appears that the basic mechanisms of recombination have been conservative in evolution. Thus the mechanisms can be investigated, by a wide variety of means, by analysing the manifestations and properties of recombination wherever and however they are revealed. However, to do so requires consideration for each case of the reproductive context and means of analysis and so separate treatment.

Homologous DNA molecules, or chromosomes, given propinquity in an appropriate biological environment, will interact to produce recombinant versions of themselves. They do so by annealing of complementary polynucleotide chains contributed by each parental molecule to form local hybrid sections and the complexes of homologues can be resolved to yield intact molecules or chromosomes. In the course of resolution, a switch of parentage may occur at or near the hybrid section or the original parental molecules may be reformed substantially. In either case, the hybrid section is usually, if not always, retained in the molecules that emerge from the interaction. The hybrid section, at least, is subject to repair which involves local destruction and resynthesis of DNA. These events may occur *pari passu* with resolution. Recombination is an orderly sequence of events conducted, like other biological processes, under the supervision of a battery of enzymes, themselves determined by genes and subject to regulation. In higher organisms, the regulation extends to restriction on the extent of possible recombination by a mechanism, the synaptinemal complex, which acts both to bring homologues into proximity and to prevent all but a minor fraction of each from interacting at the molecular level.

Molecular interaction ranges from the minimum of a fragment of single stranded DNA, contributed by a donor molecule to a competent recipient in bacterial transformation, to the maximum in which the interaction of parts of two complete molecules results in the formation of one or two molecules of joint parentage. The latter is best exemplified in the classical reciprocal crossing over and the conversion as seen in eucaryotes. Procaryotes provide examples over virtually the whole range.

This book attempts to describe the different aspects of recombination exhibited by various organisms and to relate the evidence each provides to the formulation of a revised theory of the mechanism. The areas of conflicting information about the processes, most marked among lower eucaryotes, are currently those most worthy of the effort to resolve.

The gene symbols used in practice for different organisms differ in a number of respects, most of which are relatively trivial. These include the use of numbers or letters to distinguish loci with names otherwise similar (e.g. *rec1*, *rec2* in yeast and *recA*, *recB* in *Escherichia coli*) and the presence or absence of a hyphen between locus name and number (e.g. *his-1* in *Neurospora crassa* and *his1* in yeast). In maize, it has been a frequent practice to place numbers, distinguishing different loci, as subscripts. However, it seems more convenient to write these on the line as *gl7*, rather than gl_7, for *glossy7*. It is usual to identify the 'wild' or common type allele at a locus by a superscript '+' (e.g. $his\text{-}1^+$ or $his1^+$); the wild type may be that of an arbitrarily chosen reference strain. Other mutant alleles are distinguished by letters or numbers or combinations thereof. In *Drosophila melanogaster* and maize these are written as superscripts (e.g. ry^2, wx^{90} and wx^{Coe}) but in most other organisms are written on the same line (e.g. *arg4-1* and *his1-7* in yeast). It would be defensible to enforce uniformity within the covers of one book, but this could cause trouble in going from the book to original literature. In consequence, it has been deemed wise to adhere to the usage for each organism as generally applied. In one respect only has a common practice been avoided. The book does not use a superscript '−' to designate mutants which lack a function present in the wild type; this is because the superscript '+' signifies the wild type rather than the presence of a function. The usage of '−' has arisen by confusion between gene symbol and phenotypic expression.

Acknowledgments

I gratefully acknowledge a Visiting Fellowship held in the Research School of Biological Sciences, Australian National University, during the tenure of which this book was composed and work on the control of recombination furthered. I owe very much to the encouragement of Professor Bernard John. Others have advised on particular sections, notably Professors W. Hayes and C. J. Driscoll. However, the treatment, opinions and conclusions are wholly my own responsibility. I am greatly indebted to Mrs Erica Lockwood who kindly typed the whole book and to Miss Cathy Porter who drew the illustrations. The book is greatly enhanced by the photographs, for which I am grateful to Professor John and Professor Ditter von Wettstein and to Miss Diana Combes, who helped with Fig. 1.3.

Adelaide 1976 D.G.C.

Contents

Introduction

Genetics is autonomous and must not be mixed up with physico-chemical conceptions.
M. Delbrück 1935.

Segregation and recombination are the tools of genetic analysis. Their occurrence in any organism constitutes evidence of sexual reproduction in that organism. The observation of recombination in bacteria and their viruses provided the first evidence of sexuality in these procaryotes.

Mendel was the first to enunciate these two properties of organisms. His first law stated that different heritable factors concerned with a given character segregate from one another in the formation of gametes and that in doing so each is unaltered by the other. These alternative genetic factors were later called allelomorphs or allelic genes. Mendel's second law stated that the segregation of different pairs of factors is independent, so that all combinations of non-allelic genes will be formed with equal frequency.

Within a few years of the rediscovery at the beginning of the twentieth century of Mendel's work and the confirmation of the general applicability of the laws to a wide range of plants and animals, exceptions to the second law were encountered. Parental combinations of non-allelic genes were, in some cases, found to be commoner among the gametes than were non-parental combinations. The latter ranged in frequency, according to the particular genes concerned, from nearly zero to nearly a half of the gametic output. All genes which show linkage in segregation belong to a linkage group within which they can be arranged unambiguously in a strict linear order. Genes which belong to different linkage groups commonly show independent segregation. Those in the same linkage group may show apparent independent segregation if they are far apart in the array, but will each show linkage to intermediate genes.

The number of linkage groups is characteristic of the species, being one in *Escherichia coli* and its viruses (such as T4 phage), four in *Drosophila melanogaster*, seven in *Pisum sativum* and ten in *Zea mays*. Except in special systems (see Rees and Jones, 1977), the number of linkage groups corresponds to the number of different chromosomes, the haploid number in each species. The genes of each linkage group are specified by segments of the DNA constituting the chromosomes. Recombination between linked genes occurs when homologous chromosomes come together and pair intimately at one stage of the sexual cycle. The process involved in the exchange of information between the paired chromosomes is the subject of this book. While it has many derivative functions in fertility, physiology and evolution, its basic function lies in successfully closing the cycle of sexual reproduction. It may be assumed that because of its central role in sexual reproduction and its ubiquity, the mechanism of recombination is evolutionarily conservative. Hence the mechanism may be analysed by considering the manifestations of recombination wherever and however they can be observed (review: Emerson, 1969).

At first the event could be described in simple terms as one reciprocal exchange between two homologous chromosomes occurring with similar probability at the junction between any two gene loci. Measurements of frequency could therefore be expressed as genetic maps in which the successive loci were ordered at distances approximately proportional to the recombination between them. It was shown fairly early, making use of attached-X stocks in *Drosophila melanogaster* and of trisomics in *Zea mays*, that the event of crossing over occurred at the four strand stage of meiosis, between a chromatid of each of two homologues each consisting of two chromatids.

Advance in understanding the mechanism of recombination has come from two main sources and a few subsidiary ones. One main source was the close study of events when all products of each meiosis can be recovered and the relationship between the products analysed in detail.

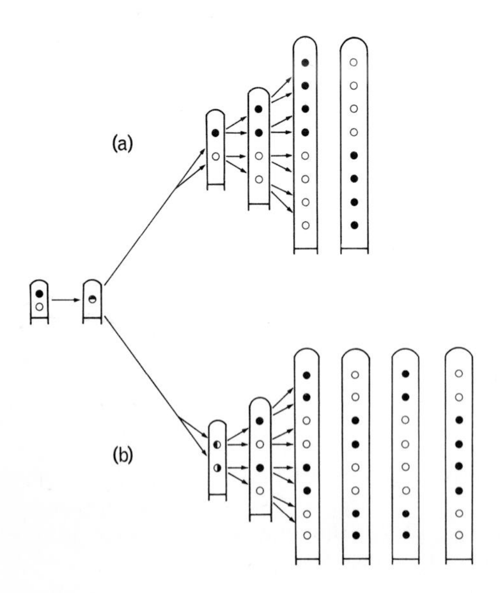

Fig. 1.1 Diagram of segregation in ordered asci: (a) at division I; (b) at division II.

A few species of fungi are the chief contributors. The other main source has been the study, especially in bacteria and their phages, of mutants which affect recombination. Behind both, in interpretation, was the realization that each chromatid is a double stranded structure, a molecule of DNA. Therefore purely mechanical break and rejoin, or even copy choice, theories were defective and had to give way to biochemical systems in which molecular events were catalysed by enzymes.

In some species of fungi, such as *Neurospora crassa* and *Sordaria fimicola*, the products of each meiosis are retained within a linear ascus in such a way that, subject to accidents, products each from one of the DNA strands of the homologues are arranged in a linear order. Counting the spores as 1 to 8 from apex to base of the ascus, the plane of the first division of meiosis lies between 4 and 5, while the planes of the second division lie between 2 and 3 and between 6 and 7 (Figs 1.1 and 1.2).

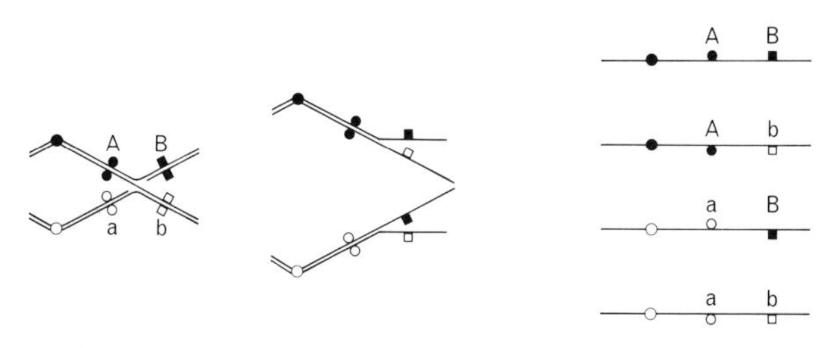

Fig. 1.2 Diagram to show relation of first and second division segregation of loci to a chiasma (or cross over). The centromeres and loci proximal to the chiasma show first division segregation, while distal loci show second division segregation. A and a are alleles at a proximal locus; B and b are alleles at a distal locus.

Each pair of spores (1 and 2, 3 and 4, 5 and 6, 7 and 8), which are sisters by a post-meiotic mitosis, represents at each homologous point the information carried by the two chains of DNA of a given chromatid as it goes through the later stages of meiosis. Segregation for one factor difference (say $+ \times m$) normally shows six types of ascus (Fig. 1.3), two with first and four with second division segregation of the factor followed (Table 1.1), all asci showing a 4:4 ratio. The frequency of second division segregation is characteristic of the particular locus, ranging from zero for those adjacent to the centromere to about 67% for those remote from the centromere. Moreover, the two first division classes are equal in freq- uency and the four second division classes are equal.. These equalities hold unless there is some genetic cause which biases the orientation of bivalents at meiosis and so makes the direction of segregation non- random, as in *Bombardia lunata* (Catcheside, 1944) where the segregation of a factor (*rubiginosa*) affecting ascospore colour can be followed (Table 1.1).

When regular segregation at more than one locus is followed in ordered asci, precise information is obtainable with respect to linkage and possible

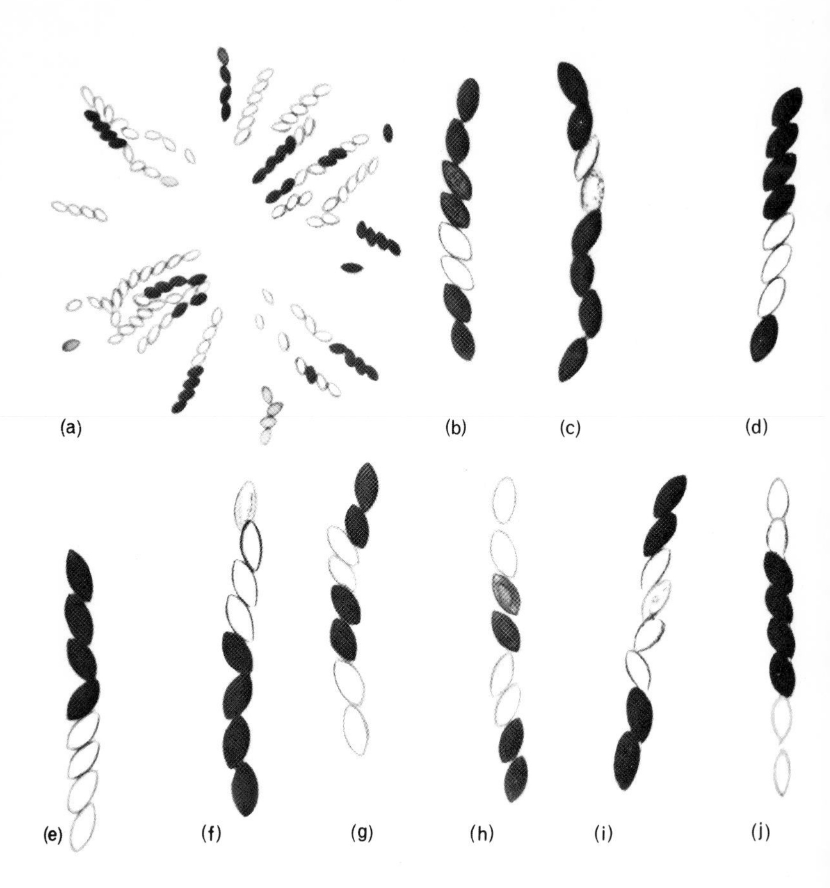

Fig. 1.3 *Neurospora crassa* asci from the cross *lysine-5* × +. The mutant *lysine-5* interferes with maturation of ascospores as well as causing a requirement for lysine; ascospores carrying *lysine-5* appear pale in asci which contain dark ripe normal ascospores. (a) is a cluster of asci. (e)–(j) are the six types shown by normal 4+ : 4*lys-5* asci. (b)–(d) are examples of abnormal asci, respectively two that are 6+ : 2*lys-5* and one that is 5+ : 3*lys-5*, presumably due to conversion of *lys-5* to normal. Apex of each ascus is above.

interference in crossing over. For two loci, there are 36 possible arrangements which may be grouped into seven modes (Table 1.2), according to whether segregation at each locus is first or second division and whether genes are recombined. The numbers of arrangements and modes increase rapidly with larger numbers of loci, to 32 modes for three loci and 172 modes for four loci. Considering the case of two loci, linkage is shown by the marked inequality of modes 1 and 2 in which both loci show first division segregation (Table 1.2). With two linked loci interference

Table 1.1 The six normal classes of segregation for a one factor difference in an ordered ascus, with illustrative data for the *mating type* locus of *Neurospora crassa* (Lindegren, 1932) and the *rubiginosa* locus ($r \equiv$ a) of *Bombardia lunata* (Zickler, 1934).

Spore	Segregation					
	Division I		Division II			
1	A	a	A	a	A	a
2	A	a	A	a	A	a
3	A	a	a	A	a	A
4	A	a	a	A	a	A
5	a	A	A	a	a	A
6	a	A	A	a	a	A
7	a	A	a	A	A	a
8	a	A	a	A	A	a
N. crassa mt	105	129	9	5	10	16
B. lunata r	1687	1271	1518	1246	1174	1308

Table 1.2 The seven modes of normal segregation observed for two loci in ordered asci. The cross is AB × ab. Change of order between planes of first (I) and or second (II) division give rise to equivalent permutations.

Mode	1	2	3	4	5	6	7
	AB	Ab	AB	AB	AB	Ab	AB
II →	AB	Ab	AB	AB	AB	Ab	AB
	AB	Ab	Ab	aB	ab	aB	ab
I →	AB	Ab	Ab	aB	ab	aB	ab
	ab	aB	aB	Ab	AB	Ab	Ab
II →	ab	aB	aB	Ab	AB	Ab	Ab
	ab	aB	ab	ab	ab	aB	aB
	ab	aB	ab	ab	ab	aB	aB
Permutations	2	2	8	8	4	4	8
Unlinked	$\frac{(1-x)(1-y)}{2}$	$\frac{(1-x)(1-y)}{2}$	$(1-x)y$	$x(1-y)$	$\frac{xy}{4}$	$\frac{xy}{4}$	$\frac{xy}{2}$
Linked opposite sides of centromere	$(1-x)(1-y)$	0	$(1-x)y$	$x(1-y)$	$\frac{xy}{4}$	$\frac{xy}{4}$	$\frac{xy}{2}$
Linked same side of centromere	$(1-x)(1-y+x)$	0	$(1-x)(y-x)$	$\frac{x(y-x)}{2}$	$x(1-y+x)$	0	$\frac{x(y-x)}{2}$

Note: The frequencies of the seven modes are given to a first approximation, neglecting double or more complex cross overs in all segments, in terms of the frequency of second division segregation of each locus, being x for Aa and y for Bb.

between two adjacent segments may be studied since the centromeres, which always segregate at division I, provide a third marked point.

Other fungi, such as yeast and *Ascobolus immersus*, keep the products of meiosis together in an unordered way. Thus with one locus segregating only one arrangement, 2A:2a or 4A:4a, is detected. With two loci three arrangements are detected. These are: 2AB:2ab, the parental ditype, comprising modes 1 and 5; 2Ab:2aB, the non-parental ditype, comprising modes 2 and 6; and 1AB:1ab:1Ab:1aB, the tetratype, comprising modes 3, 4 and 7. However, order can be imposed on such unordered tetrads or octads by means of genetic loci that show no recombination with the centromeres of the chromosomes in which they lie.

Among the first fruits of close study of *Neurospora crassa* was the confirmation of reciprocity in crossing over involving two out of four chromatids at any one place, with two, three or four strand relations showing in multiple cross overs. These occurred approximately in the 1:2:1 ratio predicted by chance coincidence of different events equally likely to involve any pair of non-sister chromatids. Extensive results from *Neurospora crassa*, summarized by Bole-Gowda, Perkins and Strickland (1962), show 423 two strand, 759 three strand and 329 four strand double cross overs as the pooled totals of a large number of studies. This represents a significant excess of two strand doubles over four stranded ones. However, other data obtained by Knapp and Möller (1955) with *Sphaerocarpus donnellii*, Strickland (1958) with *Aspergillus nidulans* and Ebersold and Levine (1959) with *Chlamydomonas reinhardi* do not show chromatid interference, though in some cases there is an excess of two stranded doubles.

In a small proportion of asci abnormal segregations are encountered in respect of a single factor difference, most being departures from the normal 4:4 segregation. These are usually 6+:2*m*, 2+:6*m*, 5+:3*m*, 3+:5*m* or abnormal 4+:4*m* (Table 1.3; some examples in Fig. 1.3). The last shows differences in respect of the locus concerned in two pairs of spores which by other criteria are sisters. The abnormalities are interpreted to arise by conversion of one gene to its allele. Lindegren (1953) followed Winkler (1932) in calling the phenomenon, expressed as departure from 4:4 or 2:2 segregation, gene conversion. Zickler (1934) was the first to prove instances of conversion. He observed 6+:2*m* in 2.46% of asci segregating at the *lactea* locus and in 0.45% at the *rubiginosa* locus, in *Bombardia lunata*. Although he also saw 2+:6*m* asci, no estimates of frequencies were reported because of possible confusion with asci containing unripe spores. However, the notion of conversion did not become accepted at all generally until later, beginning with Mitchell's (1955) work on *pyridoxin* mutants in *Neurospora crassa*. A notable feature of conversion at a locus is the correlation with crossing over in the immediate neighbourhood, making it possible to consider that the two were different consequences of one and the same event.

When recombination was found to occur at sites within a locus

Table 1.3 The normal (I) and five modes of abnormal (II–VI) segregation for a one factor difference observed in an ordered ascus, with data observed for the *gray* lcous (a = mutant, A = normal) of *Sordaria fimicola* by Kitani *et al.* (1962). The numbers of permutations of each mode take into account the various possible arrangements in an ascus; bracketed spores are sisters.

Mode	I	II	III	IV	V	VI
	{A	A	A	A	A	A
	A	A	A	A	A	A
	{A	A	a	A	A	A
	A	A	a	A	a	a
	{a	A	a	A	a	A
	a	A	a	a	a	a
	{a	a	a	a	a	a
	a	a	a	a	a	a
Permutations	6	4	4	24	24	48
Observed (i)	—	98	13	118	20	—
(ii)	—	←———	———	141 ———	———→	9
Frequencies inferred per 10^4	—	5	0.8	6	1	0.7

(Mitchell, 1955; Roman, 1956), the events were usually not reciprocal and also often required two or three classical exchanges over a very short region to account for some of the products. Examples of this sort gave rise to explanations of negative interference, one event increasing the chance of another close by. The non-reciprocal events showing, in a tetrad, as $1m^1{:}2m^2{:}1+$ for example, are more readily interpreted as gene conversion. They are the result of conversion of one mutant gene to wild type due to interaction with the allelic gene, rather than purely the result of a physical exchange. The occurrence of recombination between different sites within a locus allowed the construction, using frequencies of recombinants, of fine structure maps, an undertaking most fully pursued for the rII loci in T4 bacteriophage (Benzer, 1961). Moreover, at the molecular level, recombination may occur between adjacent nucleotides. This was demonstrated first by Yanofsky (1963) in *Escherichia coli*, using mutants causing different substitutions of one normal amino acid in the A polypeptide of tryptophan synthetase.

Theories of recombination began with ideas of breakage of non-sister chromatids at precisely corresponding sites followed by reunion in a new way. Apart from the problem of explaining how the exchanges occur with absolute precision so that unequal recombination does not occur, this mechanism does not account for non-reciprocal recombination. It was made to account for the occurrence of apparent multiple recombination in a short segment (negative interference) by supposing that pairing was intermittent, as indeed it must be, and that in paired regions exchanges could be clustered.

Lederberg (1955) suggested that recombination in bacteria might be related to the process of replication of the hereditary material, which might be copied first from one parent and then, further along, from the other parent. This copy choice hypothesis was a revival of Belling's (1928) hypothesis of crossing over. Freese (1957) applied the idea to allelic recombination in *Neurospora crassa*. The hypothesis assumed that replication was conservative so that, by copying the hereditary material of a parental chromosome so as to leave the latter intact, a process of switching between two homologues might operate when these were close together. Some lack of synchrony between the synthesis of two daughters might allow both to copy from the same parent over a short interval, so generating 3:1 ratios. Repeated switching would account for multiple exchanges. This hypothesis suffers from serious faults. For example it takes no account of the evidence that both DNA (Meselson and Stahl, 1958) and chromosomes (Taylor, *et al.*, 1957) replicate semiconservatively. It does not explain post-meiotic segregation, nor the inequality of 6:2 and 2:6 ratios. It does not explain why recombination occurs with molecular precision and that actual exchanges of material occur. It does not predict that different pairs of chromatids are involved, apparently at random at different places along a bivalent. Modifications to the copy choice theory can be made to accommodate these and other objections, but it becomes cumbersome, inelegant and hardly plausible.

All subsequent theories propose the formation of a hybrid overlap (as a joint molecule or a heteroduplex) between a segment of a chain from one DNA molecule and its complement from another DNA molecule. The joint molecule is held together by hydrogen bonding. The manner by which such a joint molecule is established may not be the same in all organisms, but it seems to be a common ingredient in all recombination processes. The idea that recombination involved the formation of a hybrid segment occurred independently to several people at about the same time, notably Holliday (1962, 1964), Meselson (1964), Whitehouse (1963) and Taylor *et al.* (1962). All made more or less detailed proposals. The notion has the explicit merit of assuming that the process is precise at the molecular level, so that the breaks, later inferred, appear to be between exactly corresponding nucleotides of the two parents. No other processes offer the appearance of such exactness. The hypothesis also has the result that the joint molecule may have mispaired bases, due to genetic differences between the parents. Correction of these by removal and replacement might occur either by excision of just one of the mispaired bases or by excision of a length of one strand including a mispaired base. Repair of the gaps would complete the conversion. Failure of correction of a mispaired base during the meiotic stages would lead to post-meiotic segregation, following replication of the chromosome or chromosomes that carried mispaired bases through the interphase.

Whitehouse (1965) pointed out that the five kinds of aberrant ascus observed by Kitani *et al.* (1962) in $+ \times g$ crosses of *Sordaria fimicola* are just what may be expected if a section of hybrid DNA is formed in two chromatids, provided that correction of mispairing does not always occur and that when it does it may occur in either direction, to normal or to mutant. If there were no correction of the mispairs, asci with $4+:4g$ and with half of them not in pairs would result (class VI in Table 1.3). The $5+:3g$ and $3+:5g$ ratios would result from correction at meiosis in one chromatid but not in the other, while the $6+:2g$ and $2+:6g$ would result from correction at meiosis in both chromatids in the same direction. Correction in both chromatids, but in opposite directions, would yield normal $4+:4g$ asci. The relative frequencies of the aberrant asci agree with this explanation. If the frequencies with which the hybrid DNA at the mutant site remained hybrid, became pure normal or pure mutant were x, y and z respectively the expected frequencies of the classes II to VI would be y^2, z^2, $2xy$, $2xz$ and x^2 respectively, assuming that the correction mechanism acts independently in the two chromatids. The observed frequencies are fitted, for the $+ \times g$ crosses in *Sordaria*, by $x = 0.27$, $y = 0.56$ and $z = 0.17$. The difference between y and z, the relative frequency of correction of the mispairing in opposite directions, might be related to the particular nature of the mutational change, whether a transition or transversion, or a deletion or insertion of bases, or to an extraneous difference. The post-meiotic segregation, implying the persistence of mispairing throughout meiosis until the synthetic (S) phase of the ensuing mitosis, could be due to failure of the postulated endonuclease to reach its substrate during a critical period of the meiotic prophase.

Whitehouse's analysis assumes that the mispaired bases in the two chromatids are alike, a consequence of his particular theory of how the segments of hybrid DNA are formed. Emerson (1966) showed that the analysis could be made more general, for the case in which the mispair in one chromatid is not like that in the other. This involves four unknown quantities, respectively p the probability of repair, with r the probability of repair to wild type, at one hybrid site, and q the probability of repair, with s the repair to wild type, at the other hybrid site. Applying the analysis to the *Sordaria fimicola* data of Kitani *et al.* (1962) and some data of Yu-Sun and himself on *Ascobolus immersus*, Emerson (1966) showed that good agreement to the data is given if $p \neq q$ and $r \neq s$ (Table 1.4). On the other hand, the *Ascobolus* data are not in agreement with $p = q$ and $r = s$, though this is fairly satisfactory for *Sordaria*.

For some time it was thought that conversion and crossing over could be independent processes. However, although the two types of events appear to be separately influenced by some conditions, the weight of evidence is that they are both the expression of a unitary process. Conversion is polarized, declining from more frequent conversion at one end of a locus to a lower frequency at the other (Lissouba *et al.*, 1962; Murray, 1963). So

Table 1.4 Agreement of estimates of repair before division I at two hybrid sites (p and q) and of repair to wild type at each site (r and s) with observed abnormal ascus types in *Sordaria fimicola* (data of Kitani *et al.*, 1962) and *Ascobolus immersus* (data of Yu-Sun and Emerson, quoted by Emerson, 1966). Analysis by Emerson (1966). In the *Sordaria* data, the number (*) of aberrant 4+ : 4*m* is estimated as the likely number, from data in Table 1.3, to accompany 239 asci of the other four aberrant types. The expectations (b) include aberrant 4+ : 4*m*, but expectations (a) do not.

		Sordaria fimicola					*Ascobolus immersus*		
			$p = 0.935$, $q = 0.694$, $r = 0.909$, $s = 0.43$		$p = q = 0.73$, $r = s = 0.767$			$p = 0.847$, $q = 0.997$, $r = 0.84$, $s = 0.29$	$p = q = 0.885$, $r = s = 0.556$
Ascus type	Fraction expected	Observed	Expected (a)	(b)	(a)	(b)	Observed	Expected	
6+ : 2*m*	$prqs$	98	98	100.9	101.7	98.4	257	257.2	230.4
2+ : 6*m*	$p(1-r)q(1-s)$	13	13	13.4	9.4	9.1	120	119.9	146.9
5+ : 3*m*	$pr(1-q)+(1-p)qs$	108	108	111.2	98.1	94.9	58	58	107.7
3+ : 5*m*	$p(1-r)(1-q)+(1-p)q(1-s)$	20	20	20.6	29.8	28.8	136	135.9	86.0
ab.4+ : 4*m*	$(1-p)(1-q)$	15*	—	7.9	—	22.9	—	—	—
	χ^2 (degrees of freedom)		0(3)	6.5(4)	5.7(3)	8.9(4)		<0.0003(3)	55(3)

recombination between two sites in the same locus depends on their relative separation as well as their degree of conversion. Polarity of conversion implies that the event could start from a fixed place to one side of the locus, a notion expressed originally as the 'Fixed Pairing Region' hypothesis. This is confirmed decisively by the behaviour of genes at recognition loci (Angel *et al.*, 1970) which affect the frequency and the polarity of conversion. In addition there is evidence that the nature of mutation at a site also affects conversion.

This outline of the theory of recombination is examined in detail in the following chapters. Ideally, it would be best if the information from different organisms could be combined into a continuous narrative. This is difficult to do at the present time for two main reasons. One is that the process in each class of organism requires knowledge of the contingent events attending recombination and these are different for eucaryotes and various procaryotes. The other is that the information about recombination is different according to the class of organism examined, the genetic system each has and the method of analysis applied.

In eucaryotes recombination is chiefly attendant upon meiosis which is described briefly in Chapter 5. At meiosis chiasmata are usually formed between homologous chromosomes after they have paired. A chiasma can be proved in some cases to be the consequence of a physical exchange, a crossing over, between homologues. Homologues are, in some cases, held together other than by chiasmata and then crossing over is absent. The nature and control of pairing and the genetic control of chiasma formation, as observed cytologically, as well as some aspects of the biochemistry of the early stages of meiosis, are also considered in this chapter. Cytological studies of meiosis permit the detection of crossing over but not of conversion. They also permit direct observation in favourable cases of the time and nature of action of genetic blocks in meiosis. These are difficult to study by other means because of the attendant sterility.

The data on conversion and the relation of crossing over to it are considered in detail in Chapter 2. This summarizes information derived from a number of fungi in which, as already described, all products of meiosis can be recovered and their ordered relation to one another in respect of the two meiotic divisions and the first post-meiotic division observed. The material considered is difficult both in respect of its detail and the apparent conflict between information drawn from different organisms. This chapter also considers evidence from organisms, especially *Drosophila melanogaster*, in which not all, at best two but usually only one, of the products of each meiosis can be recovered for analysis. There is agreement that conversion occurs in all organisms studied suitably. Moreover, the information is consistent with crossing over also occurring adjacently in about half of the cases of conversion. Conflicts of evidence arise when the direction of conversion is considered. In cases, such as that of *Sordaria fimicola* and *Ascobolus immersus*, where selected asci are analysed and the selection is dependent upon detection of a change in spore colour, the evidence suggests that the particular hybridity at a site in a heteroduplex influences the direction of conversion. On the other hand, extensive data gathered from unselected asci of yeast do not show this particular effect except in a few cases. The causation of these differences, if real, is quite unknown. There is agreement that conversion by repair in a heteroduplex, occurring at the first division of meiosis, leads to $3+:1m$ or $1+:3m$ (or 6:2 and 2:6 in octads) or, by failure of repair, to $5+:3m$ or $3+:5m$ segregations at a site. The frequency of post-meiotic segregation varies according to the nature of the mispairing at a site, from zero to a substantial fraction. Some evidence suggests that the former involves frame shift differences while the latter involves illegitimate base pairs.

Recent critical evidence suggests that a heteroduplex is formed either in only one of two interacting chromatids or else to unequal extents (pp. 41–46). In yeast, especially, excellent data show that events at two or more sites, simultaneously heterozygous at a locus, are correlated in behaviour, although the site towards one end of the locus predominates, indicating polarity in a primary event. In such heterozygotes ($m^1+/+m^2$), three classes of aberrant ascus are found: (i) 3:1, 1:3 or 1:3, 3:1 (coconversion); (ii) 5:3, 3:5 or 3:5, 5:3 (symmetrical post-meiotic segregation); (iii) 3:5, 3:1 or 5:3, 1:3 (polar conversion); in each segregation the first ratio represents $+:m^1$ and the second ratio $+:m^2$. They arise from a single heteroduplex region whose repair occurs respectively at both sites, at neither site or at only one site at a preferred end of the locus.

It has been noted previously that recombination occurs nearly universally in all organisms. In eucaryotes, it occurs sporadically at mitosis. The methods for the study of mitotic recombination and the observations, which show consistency with the process at meiosis, are described in Chapter 3. Among procaryotes, the genetic systems in bacteria are described in Chapter 6 and those in selected bacteriophages (T4 and lambda) are recounted in Chapter 7. In T4 bacteriophage some of the intermediates in recombination have been observed by electron microscopy.

Thought about the actual mechanism by which heteroduplexes are formed and the means by which they are repaired has given rise to a great range of theories which are described and compared in Chapter 8. To a considerable extent it has proved difficult to devise specific experiments that would differentiate between theories and lead to discarding those that are inappropriate. However, the evidence for asymmetry in the formation of heteroduplexes such that a heteroduplex covering sites of hybridity may be in only one of the four chromatids in a meiotic cell in a eucaryote restricts serious consideration to a small group of theories (pp. 152–156). Further, the experiments of Spatz and Trautner (pp. 111–113) with synthetic heteroduplexes cast doubt on the proposition that the mispaired bases themselves influence their fate at the time of repair.

There remains the biochemical genetics of recombination. It is clear that the process is not a purely mechanical one, but one in which each successive step is catalysed by an enzyme, each enzyme being under the control of a specific gene. Evidence to this effect is shown by a wide range of organisms, though precise enzymic properties are identified for only a few cases, particularly in bacteria (pp. 113–118) and T4 bacteriophage (pp. 124–126). At present, most losses of enzymic activity leading to general loss of recombination are identified as losses of exonucleases, even though others might be expected. In other cases the actual enzymes absent or defective in a mutant in which recombination is blocked are not yet known. In general it is also not possible to correlate the recombination mutants now known in different organisms, though analogies exist. For this reason, the information available is reported separately for each, especially in Chapters 4 and 5. So far, the correlation of genetic control of the various biological activities directly dependent on DNA metabolism, namely recombination at meiosis and mitosis, sensitivity to radiation and chemical damage and mutation, has not been completed for any organism, evidence for yeast (Table 4.1) being most advanced.

There is a second level of genetic control that operates locally (pp. 70–86) for which evidence is available only from certain fungi. There are local pairing genes that must be alike in the two homologues to permit the intimate pairing necessary to conversion and crossing over in the immediate neighbourhood. Secondly, there are local genes that appear to be the targets of enzymes (nucleases) necessary for the formation of heteroduplexes. Thirdly there are repressor genes that control the access of recombinases to local regions and so cause local depression of conversion and recombination. The systems controlling recombination locally are analogous to the operon system well known as a regulator of enzyme formation. So far, the genes and enzymes concerned in repair processes in the heteroduplexes encountered in recombination have not been identified.

This brief sketch serves not only to outline the approach to the exposition in this book. It also suggests how much there is yet to learn about what is the most fundamental property of genetic systems.

2

Recombination in Eucaryotes

Youth is the time . . . to be converted at a revival . . .

R. L. Stevenson, *Virginibus Puerisque*

Most observations of meiotic recombination in eucaryotes have relied upon randomly selected products from numerous meioses. For understanding the mechanism these data contribute little, though they provide phenomena such as positive interference that require explanation. Nearly all information about the products (tetrads or octads) of each meiosis has come from a few fungi in which the products are held together until maturity. Chiefly these are the Ascomycetes *Saccharomyces cerevisiae* (yeast), *Schizosaccharomyces pombe* (fission yeast), *Neurospora crassa*, *Sordaria fimicola* and *Ascobolus immersus*. It is convenient to deal with the evidence from each organism separately, partly because there are apparent and unresolved conflicts and partly that the refinements of the data differ. The data derived from yeast are particularly good, almost entirely because they consist of information from a large number of unselected tetrads that have been analysed fully. Moreover, the stocks of yeast are probably much more uniform with respect to genetic factors affecting recombination than are, for example, the stocks of *Ascobolus* derived from varied wild sources.

2.1 Neurospora crassa

This is a species of ascomycete that is haploid and has two mating types, each of which produce both female and male reproductive bodies. The life cycle is illustrated in Fig. 2.1. The female bodies (protoperithecia) are small dense spherical masses of hyphae in which an oogonium with one effective egg nucleus is formed; from each protoperithecium slender hyphae (trichogynes) grow towards male bodies of the other mating type. The male bodies are not specialized; one male nucleus is contributed to each protoperithecium by transfer to a trichogyne from a conidium (either macro- or microconidium) or a hypha. Each protoperithecium is stimulated to further development by fertilization and the two haploid nuclei (male and female in origin) divide conjugately many times to produce, within the wall of each perithecium, a large number of ascus initials. In each of these the two nuclei fuse to produce a diploid zygote

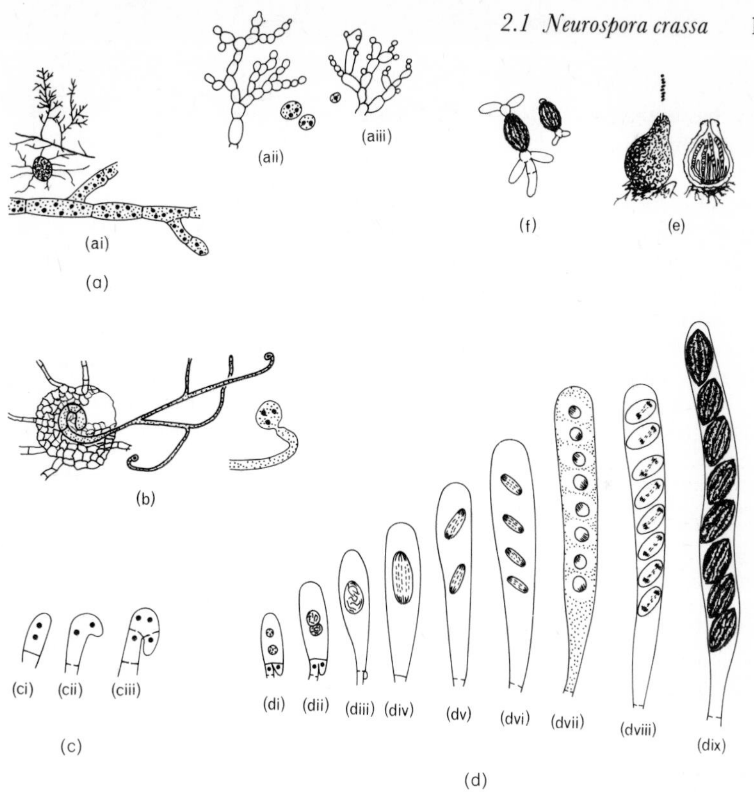

Fig. 2.1 Sketches to illustrate the life cycle of *Neurospora crassa*. In (a) is the branched multicellular mycelium (ai) consisting of multinucleate cells that connect by a small perforation in each cross wall. The mycelium bears conidiophores of two kinds, one bearing rows of macroconidia (aii) each with several nuclei, the other bearing microconidia (aiii) each with one nucleus. The mycelium also produces spherical protoperithecia, shown in detail in (b). Each protoperithecium contains an ascogonium (the female sex organ) which bears a long slender branched trichogyne. The latter grows towards and fuses with a cell or conidium, of alternative mating type, from which a nucleus passes via the trichogyne to the ascogonium. The binucleate cell produces an extensive system of ascogenous hyphae, keeping the products of the conjugate nuclei together by forming croziers (c). The terminal cell of each crozier ultimately forms an ascus (d). The nuclei enlarge (di, dii) and fuse (diii). Meiosis occurs, the two divisions (div and dv) giving a row of four nuclei, each of which undergoes a mitosis (dvi) giving a row of eight nuclei, around each of which an ascospore is delimited (dvii). Another mitosis occurs in each ascospore (dviii). The mature elliptical ascospores (dix) have a black, ribbed or 'nerved' (*Gr. neuron*) exospore. Meanwhile the protoperithecium has grown into a black, flask shaped perithecium (e) which, in section, is seen to contain many asci in all stages of development. When an ascus matures, the ascospores are forcibly shot in a row from the mouth of the perithecium. Each ascospore has a germ pore at each narrow end and hyphae are quickly formed after the ascospore is induced to germinate (f).

nucleus, the only diploid nucleus in the life cycle. Shortly the diploid nucleus undergoes meiosis to produce a linear row of four nuclei; each of which then divides again mitotically to produce a row of eight. A spore is formed around each nucleus. When the ascus is ripe it contains a linear row of ascospores that are related by descent to the divisions at meiosis. Apart from rare accidents which cause spores to be misplaced, the ordered ascus is therefore a complete record of the events at meiosis. The theory of analysis was treated earlier (pp. 2–6). Further consideration will be restricted to information bearing on the mechanism of recombination by crossing over and conversion and therefore on the analysis of events in short sections of chromosomes.

The two loci *ad-3A* and *ad-3B* are close together (recombination less than 0.5%) in the right arm of linkage group I (Fig. 2.2a). Mutation in either

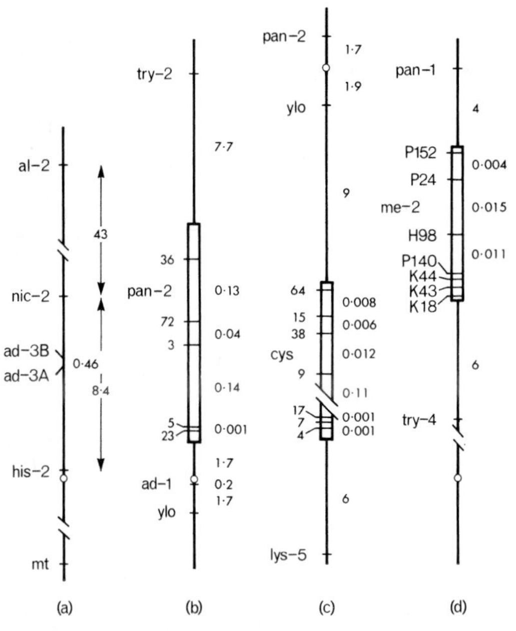

Figure 2.2 *Neurospora crassa* linkage maps of regions in which recombination in short segments is analysed in the text. The maps include magnified fine structure maps of *pan-2*, *cys* and *me-2*, in which the metric used is the percentage frequency of prototrophic recombinants. Other distances are in centimorgans.

(a) Linkage group I: *al-2, albino-2*; *nic-2, nicotinamide-2*; *ad-3B, adenine-3B*; *ad-3A, adenine-3A*; *his-2, histidine-2*; *mt, mating type*.

(b) Linkage group VI, chiefly right arm: *try-2, tryptophan-2*; *pan-2, pantothenate-2*; *ad-1, adenine-1*; *ylo, yellow*.

(c) Linkage group VI, chiefly left arm: *pan-2, pantothenate-2*; *ylo, yellow*; *cys, cysteine*; *lys-5, lysine-5*.

(d) Linkage group IV, right arm: *pan-1, pantothenate-1*; *me-2, methionine-2*; *try-4, tryptophan-4*.

locus leads to a requirement of adenine for growth and to the accumulation of a precursor that is converted into a purple pigment in the mycelium and medium. The two normal genes are not duplicates, since mutation of either leads to a requirement for adenine and *ad-3A* mutants complement *ad-3B* mutants in heterocaryons. Nor are they contiguous since deficiency for some of the chromosome between the two loci is lethal.

Giles *et al.* (1957) analysed 646 asci from the cross a *his-2 ad-3A nic-2* × A + *ad-3B* + for recombination between the two *ad-3* loci and found six asci with recombinants. The intervals *his-2*/*ad-3A* and *ad-3B*/*nic-2* could not be measured in the cross due to technical difficulties. The constitutions of the *ad-3* progeny in the six recombinant asci, recognized by each having two *ad-3*$^{+}$ spores, were determined by crosses to *ad-3A* and *ad-3B* testers and by complementation tests in heterocaryons. The six recombinant asci show two modes:

Spores								
1,2	*his-2*	*ad-3A*	*ad-3B*	+	*his-2*	*ad3A*	*ad3B*	*nic-2*
3,4	*his-2*	*ad-3A*	+	*nic-2*	*his-2*	*ad3A*	+	+
5,6	+	+	*ad-3B*	+	+	+	*ad3B*	+
7,8	+	+	+	*nic-2*	+	+	+	*nic-2*
no. of asci		5				1		

There are several significant features to note: (1) in every ascus the wild type recombinant is accompanied by a doubly mutant recombinant; (2) genes at the *his-2* and *ad-3A* loci segregate at the first division of meiosis while at *ad-3B* and *nic-2* they segregate at the second division; (3) except in one ascus where there is a further recombination (a three strand double) the *nic-2*$^{+}$ gene stays with the *ad-3B* gene. The observations are fully explained by a reciprocal cross over at a place between the loci of *ad-3A* and *ad-3B*, the exceptional ascus having in addition a cross over at a site between the loci of *ad-3B* and *nic-2*. It is unfortunate that allelic crosses at either *ad-3* locus cannot be studied satisfactorily, mainly due to their poor fertility.

Crosses involving mutants at the same locus show distinct differences from crosses between *ad-3A* and *ad-3B*. The most extensive data available for *Neurospora crassa* concern the *pan-2* locus in the right arm of linkage group VI, a map of the region with relevant loci and *pan-2* mutants being given in Fig. 2.2b (Case and Giles, 1958a, b). Two types of cross were analysed with the discovery of about one per cent of exceptional asci with recombinational events at the *pan-2* locus (Table 2.1). In two asci (A and B) reciprocal recombinants are present as in the *ad-3A* × *ad-3B* cross; in one of them a second cross over (or perhaps an error in isolation) has occurred. Reciprocal recombinants are not present in any of the other 17 asci. In all it appears as though the *pan-2* gene present in one or more spore pairs has been converted to another allele, always one for which there is information in the cross itself.

There are seven conversions of *B5* to + (asci C, D, I and O, in which there are two *B5* to + conversions), two conversions of *B3* to + (asci E,

Table 2.1 Types of abnormal asci in crosses involving *pan*-2B3 and *pan*-2B5 in *Neurospora crassa* (data of Case and Giles, 1958a; b). In setting out the exceptional asci, the mutant genes *ad-1* and *try-2* are represented by *a* and *t* respectively, while the sites of *pan-2* mutation are represented by *3* and *5*.

Cross	Constitution	Total asci	Segregation at *pan-2* locus: Normal	Segregation at *pan-2* locus: Exceptional
I	+ + 3 *try-2* / *ad-1* 5 + +	928	917	11
II	*ad-1* 5 3 *try-2* / + ++ +	846	838	8

Exceptions

Cross I							
	+ *3 t*	*a 5*+	*a 5* +	*a 5* +	*a 5* +	*a 5* +	*a 5*+
	a 53 t	*a 53 t*	*a 3 t*	*a 3 t*	*a 5* +	(*a t*)	*a 3* +
	+ + +	+ + +	+ + +	+ + *t*	{+ *3 t* / + + *t*}	+ *5* +	+ *3 t*
	(*a 5*+)	+ *3 t*	+ *3 t*	+ *3* +	+ *3 t*	+ + *t*	+ *3 t*
No. of asci	1	1	3	1	1	1	3
Reference letter	A	B	C	D	E	F	G

Cross II								
	a 53 t	*a 53 t*	*a 53 t*	*a 53 t*	*a 53 t*	*a 53 t*	*a 53 t*	*a 3 t*
	a+ *t*	*a 3 t*	*a 53 t*	*a 53*+	+ *53 t*	*a* + +	*a*+ *t*	*a 3 t*
	+ *3*+	+ + +	+ *3*+	+ *3 t*	*a 3*+	+ + *t*	+ + +	+ + +
	+ + +	+ + +	+ + +	+ + +	+ + +	+ + +	+ + +	+ + +
No. of asci	1	1	1	1	1	1	1	1
Reference letter	H	I	J	K	L	M	N	O

also showing post meiotic segregation, and F), three conversions of + to *B3* (asci J, K and L) and six asci having co-conversion of both sites. In the G asci the *B5* site is converted to + and a + site in the same gene to *B3*. In the H ascus a *B5B3* gene is converted to + while a + gene is converted to *B3*. In the M and N asci a *B5B3* gene is converted to +. The data are too few to draw any firm conclusions about the relative frequency of the various events, beyond the significant point that a third of the events involve both sites. Thus the events leading to conversion often encompass more than a third of the locus. Moreover, the two sites in the same gene are altered together, thus *B5 B3* to ++ and *B5*+ to +*B3*.

Another significant feature is that eight of the 17 asci show reciprocal recombination between the flankers *ad-1* and *try-2*, the latter showing second division segregation in seven of the asci. The frequency of recombination of *ad* and *try* correlated with conversion at the *pan-2* locus is much greater, about three times, than would be expected by chance. Five (C, D and F) of the seven asci that show second division segregation of *try-2* also show second division segregation for the elements of the *pan-2* locus that can be followed. Thus suggests that events leading to conversion in the *pan-2* locus and to correlated recombination between *ad-1* and *try-2* are initiated at a locus between the centromere and *pan-2*.

These conclusions are reinforced by data from a cross in which three sites in the *pan-2* locus were followed (Case and Giles, 1964). The data, summarized in Table 2.2, comprise 13 exceptional asci in a total of 1457.

Table 2.2 Types of exceptional asci in *pan-2* crosses in *Neurospora crassa* involving three mutant sites (data of Case and Giles, 1964). Conventions as in Table 2.1.

Constitution of cross	Total asci	Segregation at *pan-2* locus Normal	Abnormal
ad-1 B23 + B36 + / + + B72 + *try-2*	1457	1444	13

Exceptions				
	a 23 + 36 + *a* 23 72 + *t* + + + 36 + + + 72 + *t*	*a* 23 + 36 + *a* + + 36 + + + 72 + *t* + + 72 + *t*	*a* 23 + 36 + *a* + 72 + *t* + + 72 + *t* + + + 36 +	*a* 23 + 36 + *a* 23 + 36 *t* + + 72 36 + + + 72 + *t*
No. of asci	2	3	1	1
Ref. letter	A	B	C	D
	a 23 + 36 + *a* + 72 + + + + 72 + *t* + + 72 + *t*	*a* 23 + 36 + *a* + 72 + *t* + + 72 + + + + 72 + *t*	*a* 23 + 36 + *a* 23 72 + + + 23 + 36 *t* + + 72 + *t*	*a* 23 + 36 + *a* 23 + 36 *t* + 23 + + + + + 72 + *t*
No. of asci	1	1	1	1
Ref. letter	E	F	G	H
	a 23 + 36 *t* *a* + + 36 + {+ + + 36 *t* {+ + 72 + *t* + + 72 + +	*a* 23 + 36 + *a* + 72 + + {+ + 72 + *t* {+ + 72 36 *t* + 23 72 36 *t*		
No. of asci	1	1		
Ref. letter	I	J		

Two asci (I and J), like one (E) recorded in Table 2.1, show 5:3 or 3:5 segregations for the same *pan-2* sites, indicating the occurrence of segregation at the third, mitotic, division in the ascus. Only two asci (A) show the presence of reciprocal recombination within *pan-2*. All others show conversion of one or more sites: (1) conversion of 23 to + in B and C; (2) conversion of + to 23 in G; (3) conversion of + to 36 in D; (4) co-conversion of 23 and 72 in H; (5) co-conversion of 72 and 36 in I; and (6) co-conversion of 23, 72 and 36 in E and F.

As a whole, these data—the best of their time—tend to indicate a more frequent conversion from mutant to wild type. This may be valid or may be a reflection of the greater ease of detection of this type of change. The enormous labour involved in these experiments must be realized. Apart from careful dissection and careful growth of nearly 18 000 individual progeny separately, their testing by labour saving methods, such as replica plating, is not possible. In the case of the three site cross, reliance in initial classification was placed on the slight growth shown by B72 on minimal medium. However, the reports state that all progeny were tested by heterocaryon and other tests. A majority of wild to mutant conversions at the various sites occurred in co-conversions of two or three sites. Taking all the data at face value there is an indication of a decline distally in the frequency of conversion at a site from a high value at the most proximal sites (Table 2.3). But the data are insufficient to establish this firmly.

Table 2.3 Frequencies of conversions at sites in *pan-2* locus of *Neurospora crassa*. From data of Case and Giles (1958a, b; 1964) quoted in Tables 2.1 and 2.2.

Site	Conversions		Frequency
	$m \rightarrow +$	$+ \rightarrow m$	%
23	7	2	0.62
5	13	0	0.73
3	4	6	0.56
72	2	3	0.34
36	2	3	0.34

Abnormal segregations were also seen at the *ad-1* and *try-2* loci, namely two *6ad-1*:2+, two *2ad-1*:6+, one *6try-2*:2+ and two *2try-2*:6+ (Case and Giles, 1958b).

A remaining substantial series (Stadler and Towe, 1963) concerns events at the *cys* locus in VIL (Fig. 2.2c). In crosses between two *cys* alleles 45 asci showing wild type spores were classified for the site at which conversion to wild type had occurred and also for segregation of the flanking markers. About half of the asci (22 of 45) showing conversion of a *cys* site also showed recombination between *lys-5* and *ylo* such that *lys-5* showed second division segregation in these asci (Table 2.4). However, only four of these asci showed second division segregation at the *cys* locus, showing that the site of initiation of conversion at the *cys* locus and of correlated recombination with *lys-5* in half of the events lies distally to *cys* between it and *lys-5*. The data do not clearly show any evidence of polarity in conversion.

Although conversion was first demonstrated unequivocally at the *pdx* locus in this species (Mitchell, 1955) the analysis of unselected asci has been much less extensive than for yeast. No doubt the greater labour and the rather low frequency of conversion has been a deterrent. Observed values are: 0.7% at *pdx* (4/585), 1% at *pan-2* (28+4 c.o./3178), 1.7% at *cys* (14/1651) based on cys^+ only). If the maximum observed frequencies of prototrophic spores observed in allelic crosses are used as a guide and it is assumed that an equivalent proportion of doubly mutant auxotrophs occur, the estimated frequencies at a range of asci would be: *am-1* 0.9%; *cys* 1.2%; *his-1* 0.4%; *his-2* 0.8%; *his-3* 2%; *his-5* 0.07%; *me-2* 0.8%; *pan-2* 0.6%.

Table 2.4 *Neurospora crassa* asci, from crosses of the type *lys-5* cys^x + × + cys^y *ylo*, that show conversion of a *cys* mutant site to cys^+. Data of Stadler and Towe, 1963. The P asci have *lys-5* and *ylo* in parental combinations; R have *lys-5* and *ylo* recombined in two spore pairs. In every case the proximal locus (*ylo*) showed division I segregation; in every R case the distal locus (*lys-5*) showed division II segregation.

	Conversion $cys^x \rightarrow +$		Conversion $cys^y \rightarrow +$	
$cys^x \times cys^y$	P	R	P	R
17 × 64	9	7	7	5
4 × 38	2	3	1	0
38 × 64	1	1	1	2
9 × 64	1	2	1	2
Totals	13	13	10	9

In summary, the data from *Neurospora crassa* show that: (1) events of recombination within a locus are usually not reciprocal; (2) about half of the events are accompanied by reciprocal recombination between flanking markers; (3) this is chiefly due to crossing over between the converted locus and one flanker; (4) events involving two or more sites simultaneously in a locus (co-conversion) are common; (5) the events are polarized as though initiated from a particular end of the locus; (6) the point of initiation is proximal to *pan-2*, so that crossing over occurs between *pan-2* and *ylo*; (7) the point of initiation is distal to *cys*, so that crossing over occurs between *cys* and *lys-5*; (8) conversion may occur in either direction, from mutant to wild type or the reverse. Further tests of these conclusions and of the relative frequencies of different kinds of events are dependent upon consideration of the much more extensive data obtained in yeast.

However, data bearing particularly on polarity has been obtained for several loci in *Neurospora crassa* using data from random spores. The method is to isolate m^+ recombinants from crosses of the type

$Pm^1+D \times p+m^2d$, where m^1 and m^2 are alleles, mutant at different sites, and P, p and D, d are proximal and distal markers as close either side of m as is possible. Events may in different cases be initiated at a locus, outside m, either proximal or distal to the m locus (Fig. 2.3). It is presumed that conversion will occur more frequently, with a relative frequency x (greater than 0.5), at m sites closer to the initiation locus. If recombination occurs between p and m or between m and d, depending on the site of initiation, in a fraction y of the cases the quantitative relations between the four classes, two parental ($P+D$ and $p+d$) and two recombinant ($p+D$ and $P+d$) will be as in Fig. 2.3 and Table 2.5. With a proximal locus of initiation, the major parental class is $P+D$ and the

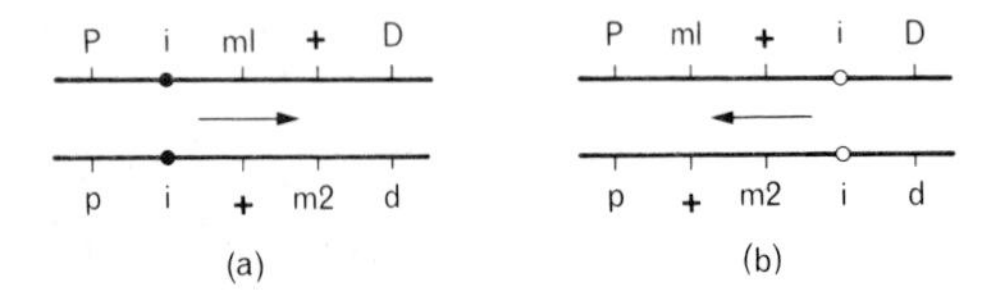

Fig. 2.3 Diagrams to illustrate conversion promoted from a proximal (a) or distal (b) initiation site (i). Conversion to wild type is more frequent ($x > 0.5$) at the site (*m*1 or *m*2) in the *m* locus nearer to the initiation locus. Recombination between the *m* locus and either p or d occurs near the initiation locus in a fraction y. The quantitative effects on the four classes of prototroph, $P+D$, $p+d$, $p+D$ and $P+d$ are given in Table 2.5.

Table 2.5 Frequencies of the four classes of prototroph derived from proximal or distal initiation of a heteroduplex (see Fig. 2.3 above) leading to more frequent conversion proximally or distally respectively. The higher conversion frequency ($x > 0.5$) and recombination between flanking markers (y) lead to inequalities of the four classes.

	Constitution of prototrophs			
Site of initiation	$P+D$	$p+d$	$p+D$	$P+d$
proximal	$x(1-y)$	$(1-x)(1-y)$	xy	$(1-x)y$
distal	$(1-x)(1-y)$	$x(1-y)$	xy	$(1-x)y$

major recombinant class $p+D$. With a distal locus of initiation, the major parental class is $p+d$ and the major recombinant class $p+D$. The map order of the sites in the m locus is indicated primarily by the predominance of the $p+D$ recombinant class and secondarily by the predominance of one parental class, provided that conversion in the m locus is polarized. The direction of polarity is indicated by the ratio of $P+D$ to $p+d$, which is greater than unity for proximal and less than unity for distal loci of initiation. Examples of these data are given in

Table 2.6. It is surprising that in five of the seven examples the initiation locus appears to be distal to the locus in which conversion is occurring; the *his-1* and *am-1* loci are the exceptions. Curiously, these data for the *pan-2* locus appear to be in conflict with the indications from complete asci. The direction of polarization is not affected by removal to another chromosome (Murray, 1968) in the same order with respect to the centromere (*me-2*) or in a reverse order (*me-6*). Generally the frequency of recombination of the flankers is about 50%; but *me-7* and *am-1* are exceptional in having low values.

Table 2.6 Data on polarity and associated crossing over drawn from random prototrophic recombinants in allelic crosses. Indices are of polarity (x) and crossing over between proximal and distal flankers (y). Polarity indices are in bold type for those of proximal origin, all others being of distal origin. Sources of data: (1) Thomas and Catcheside, 1969, (a) *rec-1* × *rec-1*, (b) *rec-1*$^+$ *rec-1*; (2) Smith, 1965; (3) Stadler and Towe, 1963; (4) Smyth, 1971; (5) Case and Giles, 1958b; (6) Murray, 1969; (7) Siddiqui, 1962; (8) Nelson, 1962.

Locus and alleles m^1, m^2			P/p	D/d	PD	pd	pD	Pd	x	y	Reference
Neurospora crassa											
his-1	K83	K625	*am-1*/+	+/inos	2392	1585	2150	1138	**.62**	.45	1a
					998	1827	1331	719	.65	.42	1b
his-5	K516	K548	+/*pyr3*	*leu-2*/+	24	145	219	19	.89	.52	2
cys	9	17	*ylo*/+	+/*lys-5*	44	109	100	44	.70	.49	3
am-1	K314	*am-1*^{11}	*sp*/+	*his-1*/+	3334	476	808	662	**.74**	.37	4
pan-2	B5	B3	*ad-1*/+	+/*try-2*	30	82	85	15	.80	.42	5
me-2	K18	K44	*try-4*/+	+/*pan-2*	83	43	155	34	**.75**	.55	6
	K18	P140			59	41	141	33	**.68**	.54	
	K18	H98			46	91	74	18	.74	.36	
	K43	K44			54	28	94	16	**.77**	.50	
	K43	P140			42	49	82	18	.70	.46	
	K44	P140			12	50	38	18	.75	.52	
	K44	P152			16	83	32	17	.76	.39	
	H98	P152			19	75	35	27	.69	.45	
	P24	P152			44	117	58	39	.67	.40	
me-7	250	73	*thi-3*/+	*wc*/+	115	282	89	40	.70	.25	6
	250	254			109	101	25	9	**.63**	.13	
	254	73			14	123	12	7	.80	.18	
Aspergillus nidulans											
pab-1	5	6	+/*ad9*	+/*y*	18	53	107	9	.86	.50	7
Zea mays											
wx	90	Coe	+/*v*	+/*bz*	15	27	63	3	.86	.41	8

Some sets of data, those for *me-2* (Fig. 2.2d) and *me-7* especially, appear to indicate a reversal of polarity within a locus (Table 2.6). The usual polarity at the *me-2* locus is from the distal end with *pd* much greater than *PD*. This reversal has been ascribed to preferential conversion of the sites located close to the ends of the locus and as due to there being initiation points at both ends of the locus. In the case of *his-3*, as shown in Chapter 4, apparent reversal of polarity is due to heterozygosity at an initiation locus distal to the *his-3* locus. By switching the same *his-3* alleles between the two initiation genes (*cog* and cog^+) polarity can appear to be reversed. The same effects are possible for *me-2* and *me-7* for which the necessary tests for possible variation in an effective initiation locus outside them have not been made. Until such effects have been excluded, changes in direction of the polarity of allelic recombination as a function of the position of the sites within a locus cannot be regarded as established. If a locus were subject to conversion initiated from both ends and with each predominating towards its respective end, one would expect to detect the action of one by eliminating the other, as in the case of the *rec-1*$^+$ gene acting on the *his-1* locus. However, although there is a large reduction of *his-1*$^+$ prototrophs, to a fifteenth, it may be calculated (Thomas and Catcheside, 1969) that the proportion of distally initiated conversion is not greater than 9.5% in the absence of *rec-1*$^+$. The small amount of conversion initiated distally would not show as a reversal of polarity.

2.2 Yeast (Saccharomyces cerevisiae)

The life cycle is illustrated in Fig. 2.4. There are two mating types, *a* and α, and crosses are readily effected by pairing haploid vegetative cells or ascospores of different mating type. The resulting diploid cells divide mitotically unless induced to enter meiosis. Hence recombination at

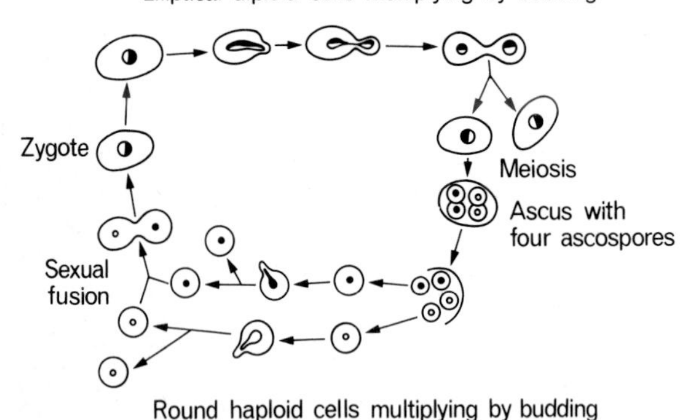

Fig. 2.4 Life cycle of yeast (*Saccharomyces cerevisiae*).

mitosis and meiosis can be studied in the same hybrid. By appropriate procedures, it is possible to obtain tetraploid or triploid strains and also strains which are disomic for one chromosome and monosomic for all others. These are useful for several purposes, but the discussion of recombination will be restricted to events in diploids. If induced by starvation in a suitable medium (free of nitrogen and providing acetate as the only carbon source), each diploid cell enters on meiosis. This results in four haploid nuclei around each of which an ascospore is formed, the original cell being converted into an ascus. The four spores are readily isolated by micromanipulation and the resulting clones grown from them can be scored for known marker genes by various techniques including replica plating. Sixteen linkage groups containing over 100 genetic loci are known.

A diploid heterozygous at two loci, Aa and Bb, produces mainly three classes of asci: parental ditypes (PD) with 2AB and 2ab, non-parental ditypes (NPD) with 2Ab and 2aB and tetratypes (T) with one each of AB, ab, Ab and aB. The proportions of these three classes of asci depend on the positions of the marker loci with respect to their centromeres and whether they are on different chromosomes or the same one and then whether or not on the same side of the centromere. For A and B unlinked and close to the centromeres, PD = NPD with the magnitude of T reflecting the extent of recombination in the two gene to centromere intervals. As these intervals show increased recombination, the ratio PD:NPD:T changes towards 1:1:4 as a limit. If A and B are linked, PD asci predominate; recombinants between the A and B loci generate T and some NPD asci, the latter resulting from four strand double exchanges. Estimation of recombination between a locus and its centromere requires the identification of the first division array of spores. The sister spore pairs which shared a common pole at the first meiotic division were first identified in linear asci by Hawthorne (1955). Once loci closely linked to centromeres were known, the sisters could be identified in unordered asci (Hawthorne and Mortimer, 1960).

Loci in different linkage groups segregate independently. Positive chiasma interference is observed between adjacent regions within one arm of a chromosome, but not between regions on opposite sides of the centromere. Chromatid interference is, with rare exceptions, absent. Reports of segregations suggesting preferential directions of segregation at the first meiotic division leading to dependent segregation of unlinked genes are suspect. In cases which have been tested critically, the effect was shown to be due to selection.

In tetrads derived from single meioses, it is expected that every heterozygous site (Aa) will segregate 2A:2a. However, exceptional classes occur and are usually 3A:1a or 1A:3a. Post-meiotic segregation, in which usually one of the four ascospores proves to be heterozygous at the marked site, is also observed. These exceptions are gene conversions, unless some other genetic process (such as polyploidy, non-disjunction, suppression, mitotic recombination) is implicated.

Fogel, Hurst, Mortimer and their associates have collected data on a large number of unselected tetrads. The frequency of gene conversion varies considerably from one heterozygous site to another, e.g. the *mating type* locus converts with a frequency of 0.3% while *Sup6* (an ochre suppressor) converts in 15–20% of asci. The conversion frequency for a given heterozygous site is fairly constant from cross to cross in yeast and does not appear to be related at all to position in the linkage map. However, at a particular locus different alleles convert with different frequencies. This variation commonly shows polarity in that the frequency of conversion increases with increasing separation from an allele with a low conversion frequency. Thus the spontaneous meiotic conversion frequencies for the alleles *arg4-4*, *arg4-19*, *arg4-1*, *arg4-2*, *arg4-16* and *arg4-17*, with sites of difference in that order in the *arg4* locus, are respectively 1.6, 2.8, 5.5, 6.9, 6.2 and 7.1% (data of Hurst *et al.*, 1972). Maps may be made of a locus by measuring the frequency of prototrophs formed by spontaneous recombination at meiosis or the frequency of mitotic recombination induced in diploid vegetative cells by X-rays (Manney and Mortimer, 1964). In the latter method the frequency of induced prototrophs rises linearly with increasing dosage, the rate of rise varying with the pair of alleles used. The regression slope is used as a measure of separation. Defining 1 unit as 1 prototroph per 10^8 survivors per roentgen, the map of the *try5* locus, which is responsible for tryptophan synthetase, is about 10 units. The corresponding polypeptide consists of about 500 amino acids, so one X-ray map unit corresponds to about 50 amino acids or about 150 nucleotides. Parker and Sherman (1969) have given evidence for a revised relationship of one X-ray map unit as equivalent to 129 nucleotide pairs. The two kinds of map (prototroph frequency and X-ray) show concordance in order and relative spacing within the limits of experimental error.

Typically, the frequencies of 3A:1a and 1A:3a segregations are equal in yeast within statistical limits. Conversion is also a conservative process in that it preserves, but does not generate new, genetic information. Conversion operates with complete fidelity in that the converted and the parental genes are indistinguishable by any genetic test (suppression, recombination, gene product, complementation response, suppressibility) that is applied.

Most data on gene conversion are derived from diploids which are heterozygous in the repulsion phase at two sites within a locus, i.e. *a1+/+a2*, commonly described as heteroallelic. In these, meiotic conversion may be observed in three ways: (1) selected prototrophic spores; (2) selected asci containing a prototropic spore; (3) unselected asci. Prototrophic spores arise from 3+:1a (or 5+:3a) segregations of either allele or from reciprocal recombination (1++:1*a1 a2*:1*a1*+:1+*a2*) between the mutant alleles. Among selected prototrophic spores it is not possible to determine which event was involved. Among selected asci, the events can be recognized. Thus among 1081 tetrads selected by Fogel and Hurst (1967) on the basis that they contained a prototrophic spore, about 90% had 3:1 conversions of one or other allele and only 10% arose by an apparently reciprocal event between the mutant sites.

The data in Tables 2.7 and 2.8 (Fogel and Hurst, 1967) were derived from three crosses, involving respectively *his1-7*× *his1-1*, *his1-315*× *his1-1* and *his1-315*× *his1-204*; the order of the sites in the *his1* locus is, from the proximal end, 315, 7, 204, 1. The three crosses (Table 2.7) differ significantly in the proportion of asci showing reciprocal

Table 2.7 Recombination events in the *his1* locus of yeast. (Data of Fogel and Hurst, 1967).

Cross	Prototrophs per 10^5 asci	Reciprocals % of abnormal asci	Conversions Proximal	Conversions Distal
his1-7 × *his1-1*	293	12.6 ± 1.4	452	75
his1-315 × *his1-1*	416	7.6 ± 1.6	234	34
his1-315 × *his1-204*	344	1.6 (± 0.9)	161	24

Table 2.8 Distribution of combinations of flanking genes in *his1*$^+$ recombinant ascospores in the yeast crosses + *his1*-x + *arg6* × *thr3* + *his1*-y +, summarized in Table 2.7. (Data of Fogel and Hurst, 1967). *arg6* = *arginine 6*; *thr3* = *threonine 3*.

Type of *his1*$^+$ recombinant	+ *arg6*	*thr3* +	*thr3 arg6*	+ +
Reciprocal	4	0	97	0
Proximal convert	474	5	265	103
Distal convert	0	49	77	7
Totals	478	54	439	110

recombination within *his1* (χ^2 = 21.9 for 2 degrees of freedom), but the ratio of proximal to distal conversions is constant (Table 2.7) over the three crosses (χ^2 = 0.43 for 2 degrees of freedom). This suggests a general difference in origin of the two kinds of recombination, with conversions arising from an event proximal to *his1* whereas the reciprocals arise from an event within *his1* with different frequencies affected by the detailed structure of the particular *his1* genes. When such prototrophic spores are scored for closely linked flanking markers, one parental and one recombinant array usually predominate (Table 2.8). This is expected if one site converts more readily and it provides a basis for assessing polarity relations among various sites in a locus. In each cross the proximal *his1* allele was accompanied distally with the mutant gene *arg6*, while the distal *his1* allele was accompanied proximally with *thr3*, so that the crosses were + *his1-x* + *arg6* × *thr3* + *his1-y* +. The distribution of the four combinations of the flanking genes differs markedly between the three classes of asci (Table 2.8); the three crosses are pooled, their data in these respects being homogeneous. It may be noted that, among the converts, recombinants of the flanking markers are nearly as frequent (452) as the parentals (528). However, among the reciprocals one class of recombinant of flankers occurs almost exclusively; this is the one that indicates the order of the sites of difference in the *his1* locus.

Similar conclusions follow from the analysis of unselected asci, but additional features are disclosed. When a substantial sample of unselected asci from crosses involving alleles in yeast is analysed, seven classes of recombinant tetrads are found (Fogel and Mortimer, 1969). Besides the 3+ : 1*m* and 1+ : 3*m* for either parental allele and the reciprocal allelic recombinants, there are conversions that simultaneously involve both sites of a single parental strand. These asci with co-conversions have three spores that carry one of the parental alleles and one that carries the other. They are due to an event in which a segment of one chromosome is replaced by information equivalent to that in the corresponding segment of the homologue. This conclusion is based on the rarity of six other types of tetrads which should occur commonly if the co-conversions were due to independent events of conversion. Alleles with widely separated sites of differences usually convert independently. As the sites come closer, the frequency of co-conversion increases at the expense of single site conversions. The proportion of single site plus double site conversions is virtually constant and independent of the relative positions in the locus of the sites of difference as is illustrated for the *arg4* locus (Table 2.9). The indication is that virtually the whole of the locus is usually involved in each event leading to conversion and that the polarity of frequencies reflects departures from this.

Table 2.9 The seven types of recombinant asci observed in crosses between pairs of alleles at the *arg4* locus of yeast. Proximal allele first. (Data of Fogel and Mortimer, 1969). Order of alleles: *arg4-4, -1, -2* and *-17.*

	Alleles crossed	*4×17*	*1×2*	*2×17*
	Nucleotide separation	1060	520	128
Type of recombinant	Ascus constitution			
Conversion of one site				
proximal $m \rightarrow +$	$m^1+, +m^2, +m^2, ++$	3	3	1
$+ \rightarrow m$	$m^1+, m^1+, +m^2, m^1m^2$	5	3	3
distal $m \rightarrow +$	$m^1+, m^1+, +m^2, ++$	18	10	3
$+ \rightarrow m$	$m^1+, +m^2, +m^2, m^1m^2$	20	11	2
Co-conversion of two sites				
proximal $m \rightarrow +$, distal $+ \rightarrow m$	$m^1+, +m^2, +m^2, +m^2$	2	13	14
proximal $+ \rightarrow m$, distal $m \rightarrow +$	$m^1+, m^1+, m^1+, +m^2$	1	10	13
reciprocal	$m^1+, ++, m^1m^2, +m^2$	9	5	0
exceptional		1	1	0
Total abnormal asci		59	56	36
Total asci analysed		697	502	544
% of asci with conversion		7	10	6.6

In cases where three or four sites of difference within a locus are followed in a heterozygote, co-conversion of all sites can occur (Table 2.10). Indeed, the frequency of these events is quite high. Generally, if two sites are converted any other site of difference between them is also

Table 2.10 *Saccharomyces cerevisiae*. Conversion at the *arginine4* locus in multiple heterozygotes. (Data of Hurst *et al.*, 1972).
Order of *arginine4* sites: *19, 1, 2, 16, 17*.

arginine 4 genotype	Total asci analysed	Sites converted	No. of asci converted
19 + 17 / *+ 16 +*	2566	*19*	19
		16	16
		17	41
		19, 17	10
		19, 16	2
		16, 17	82
		19, 16, 17	42
1 2 + + / *+ + 16 17*	1505	*1*	4
		16	2
		17	15
		2, 16	4
		1, 2, 16	2
		2, 16, 17	31
		1, 2, 16, 17	74

converted; only ten exceptions to this rule (the *19 17* co-conversions) occur among 161 asci (in Table 2.10) to which it could apply. The indications are therefore that the segment of the chromosome in which correlated events of conversion can occur is rather long, the modal length being about 1000 nucleotide pairs.

An important observation is that deletions, in heterozygotes, are converted (Fink and Styles, 1974) to wild-type and the wild-type to the deletion. The two events occur with nearly equal frequency. Two deletions, *his4-15* and *his4-29*, each lacking part of the left end of *his4* have been tested in heterozygotes with a site mutation, *his4-290*, at the right end of *his4*. Conversion of the deletions occurs at rates, 5% and 2% respectively, which are less than the 8.2% for a point site (*his4-39*) in the same part of *his4*. The rate of conversion at the *his4-290* site is reduced in heterozygotes with the larger deletion (*his4-29*) to 1.6% from its usual value of 5.2%. Co-conversion also occurs. The reduced rates of conversion due to the larger deletion are probably due to imperfect pairing in the heterozygote, since in a diploid homozygous for this deletion conversion at the *his4-290* site returns to normal. There is no direct evidence available in yeast to show whether small deletions or duplications, such as would result from frame shift mutation, show equal or unequal conversion to and from wild-type. However, the behaviour of large deletions makes any difference unlikely. The question of whether different kinds of transition and transversion mutants have different effects is also unresolved. Differences are unlikely, but since the natures of the point mutations used are unknown it cannot be asserted unreservedly that no differences could arise from this source.

Gene conversion is associated with a high frequency of recombination of flanking markers. Hurst, Fogel and Mortimer (1972) have summarized their data from 11023 unselected tetrads in which there were 907 conversions at six loci on three chromosomes. 455 (49.1%) of these conversions were accompanied by recombination between the flanking markers that were within 20 map units apart. In crosses in which there were three or four heterozygous sites in a locus, 22 were found with conversion of an internal allele. Of these, 14 were recombined for flanking alleles within the same locus. It is typical for half of all conversionary events to be inside a reciprocal recombination of neighbouring material. The data strongly suggests that all crossing over originates in conversionary events.

Assuming that the loci at which conversion has been studied in yeast represents a fair sample, the average conversion frequency, about 1%, represents the fraction of all the chromosomes involved in conversion at each meiosis. Conversion of 1% of the yeast chromosomes (1.5×10^7 nucleotide pairs) would involve 1.5×10^5 nucleotide pairs. With a modal length of each converted segment estimated at 1000 nucleotide pairs, this amount of converted DNA corresponds to 150 events of conversion per meiosis, half of which (75) would be associated with crossing over. With no other source of crossing over, this number of cross overs would correspond to a total genetic length of yeast chromosomes equal to 3750 cM. This estimate agrees well with the value of 3600 centiMorgans obtained from mapping data (Mortimer and Hawthorne, 1966). Quantitatively, the observations are consistent with there being one mechanism for crossing over, associated with conversion.

2.3 Schizosaccharomyces pombe

In fission yeast similar results to those found in brewer's yeast have been reported. These include the occurrence of conversion and of co-conversion. The life cycle is similar to that in *Sacc. cervisiae*. Mutants at the *ade6* locus require adenine for growth and accumulate a red pigment if grown on media with limited amounts of adenine. Thus in a cross of two *ade6* mutants the occurrence of a convert to normal *ade6*$^+$ is detectable by its white colour. The intensity of the pigment differs between *ade6* mutants, M26 and M375 forming dark red colonies while M216 and L52 form light red colonies. In suitable crosses, e.g. M26 × M216, the occurrence of converts other than to *ade6*$^+$ are detectable. Further, post meiotic segregation can be detected as sectored colonies.

Gutz (1971b) has reported an exception to the parity of 3A:1a and 1A:3a segregations found among tetrads with conversions. This concerns the mutant M26 at the *ade6* locus. However, a number of correlated peculiarities suggest that the special properties could be due either to the *ade6-M26* gene or to another genetic factor linked to it. When M26 is present in crosses to other *ade6* alleles the frequency of prototrophic recombinants is up to 20 times greater than in crosses not including M26. Conversions are found in 3.9% of asci from crosses including M26, and only 0.6% in those without M26. In crosses of M26 to other *ade6* alleles, M26 is the more commonly converted to $ade6^+$, irrespective of whether it is proximal or distal to the site of the other allele. The other allelic site in the cross undergoes co-conversion, always if close to M26 as is M216, reducing to about 60% of asci if further away as is L52.

Gutz has postulated that M26 itself influences the frequency with which gene conversion is initiated in the *ade6* locus and that there may be a preferred breakage point at the M26 site. The case is regarded as a marker effect, with those in *Ascobolus immersus* (see page 34). However, there are obvious parallels with the action of cog^+ at the *cog* locus near to *his-3* in *Neurospora crassa* both in the raised frequency of conversion and the bias to conversion in a particular chromosome in heterozygotes (see Chapter 4). No experiments to determine whether M26 is separable from these properties have been reported.

2.4 Ascobolus immersus

Ascobolus immersus is a heterothallic ascomycete with two mating types and producing apothecia. Each ascus contains eight ascospores arranged in an unordered cluster which is discharged in a coherent group that can be caught on a suitable surface. The ascospores are large (65–70 μm) and in the wild strains are pigmented a dark purple brown colour. They germinate readily on bacto-peptone agar medium containing 1.5% NaOH after overnight incubation at 40°C. Numerous mutants, at several different loci* (*46, 19, 75, 164, 726, 231, XXVI, 84W*), are known to produce ascospores lacking pigment (pink or white) or in which the pigment is condensed in drops (rough or granular). These have proved useful for the study of conversion, relying upon the observation of coloured ascospores in asci arising from crosses between allelic colourless mutants. In crosses of the type $c^1+ \times +c^2$, the c^+ prototrophs appear to be mainly from one of the parents. Thus most asci which contain a prototroph are of the type $2c^1+$, $2++$, $4+c^2$; c^1+ is the minority parent, $+c^2$ the majority parent. If a number of alleles is studied they may be arranged in a consistent order, say c^1, c^2, c^3, c^4, etc., such that the left hand one of any pair is that which is the more frequently converted to prototrophy. The order is consistent with an order based on the frequencies of asci containing a recombinant, on the assumption that these frequencies may be used as additive mapping functions. The group of alleles which show this relation are said to constitute a polaron. Explanations advanced depend upon the suggestion that the events of recombination are polarized, commencing with strand breakage at a

* Usually referred to as 'series' rather than locus in the special literature. No good genetic map has been prepared for *Ascobolus immersus*.

fixed point at the start of a polaron and that hybridity corrections are made generally only at the site of the first allelic difference encountered. In crosses of the type $+ \times c$, abnormal segregations of the type $2+:6c$ and $6+:2c$ may be observed, as well as $3+:5c$ and $5+:3c$. Care has to be taken in interpreting those with more than four colourless spores because some spores that are genotypically + may appear uncoloured.

Allelic crosses in *Ascobolus immersus* show the occurrence of $2+:6c$ asci with frequencies ranging up to 5% or more according to the locus and the particular alleles. At any particular locus the sites of difference of the alleles can be arranged in linear order using as metrics: (1) the frequency of abnormal $2+:6c$ asci used as an additive function; (2) the site showing predominance in conversion to c^+. The two methods generally give concordant results. An example is given in Table 2.11 for locus *75*. Different loci and, indeed, different alleles at any one locus appear to show distinct and individual properties, though no experiments to separate such properties from locus or site have been attempted.

Table 2.11 Data (Lissouba *et al.*, 1962) for ordering mutant sites of five alleles at the *75* locus of *Ascobolus immersus*. Order of the sites of mutation is: *231*, *322*, *278*, *147*, *1987*.

		Analysis of samples of $2+:6c$ asci		
		Conversions		Cross overs
Alleles crossed	$2+:6c$ asci %	left	right	
231 × *322*	1.0	7	4	0
231 × *278*	1.5	9	4	1
231 × *147*	3.3	11	1	4
231 × *1987*	3.3	11	1	5
322 × *278*	1.4	7	6	3
322 × *147*	2.9	4	2	6
322 × *1987*	6.7	3	1	13
278 × *147*	1.7	12	4	3
278 × *1987*	3.5	12	3	3
147 × *1987*	2.0	8	8	1
Totals		84	34	39

Analysis of $2+:6c$ asci (Table 2.12) shows that some are the consequences of reciprocal recombination (cross overs) while others are not (conversions). The frequencies of cross overs range from less than 0.5% of the exceptional asci at the *46* locus to 25% at the *75* locus. The behaviour at locus *19* is especially interesting. The large number of alleles can be grouped into three sets A, B and C, within each of which abnormal asci are very infrequent, less than 0.02%. Frequencies are higher in crosses between members of different groups, viz. 0.16% in A × B and B × C and

Table 2.12 *Ascobolus immersus*, occurrence of conversion and crossing over in abnormal asci at various loci. Data of Lissouba *et al.*, 1962 (loci *46*, *75*, *19*); Kruszewska and Gajewski, 1967 (locus *y*); Makarewicz, 1964 (locus *726*); Baranowska, 1970 (locus *164*).

Locus	Alleles	Abnormal asci % (maxima)	2+ :6*c* asci Cross overs	2+ :6*c* asci Conversions
46		0.5–1.0	1	202
75		5–7.5	39	118
19	A×A, B×B, C×C	0.02	0	48
19	A×B	0.16	36	51
19	A×C	1.92	194	241
19	B×C	0.16	0	41
19	A60×C55		88	46
19	A60×C324		15	9
19	A1678×C58		11	36
19	A1678×C324		15	71
y	*y*×77	0.27	1	29
726		1.0	21	69
164		1.0	277	226

1.92% in A×C; the values are not additive. No cross overs occur in A×A, B×B, C×C and B×C crosses, but relatively large proportions occur in A×B and A×C crosses. However, different A alleles, for example, show significant differences from one another in crosses to common C alleles; crosses involving A1678 show 20% of cross overs, compared with 65% in crosses involving A60; A60×A1678 has shown no recombinants in over 4×10^4 asci.

The variations between and within loci suggest genetic causes, either due to the particular mutant alleles or to other factors derived from the wild-type strains which are likely to be heterogeneous. For example, there could be a site, between the A and B clusters of locus *19*, at which recombination leading to reciprocals and conversions begins, as well as a site outside to the right of locus *19* leading only to conversions in C and B. The alleles A60 and A1678 show different probabilities for the two events. The existence of two sites of initiation of recombinational events at locus *19* would account for the observed reversal of predominance of conversion in A×B compared with A×C and B×C crosses. In the former, A×B, the A site is the more commonly converted, whereas in A×C and B×C the C site is the predominant convert.

Allele 277 at locus *46* shows a significant proportion of 1+ :7*c* asci, about a quarter of the frequency of 2+ :6*c*. Its site of difference is close to that of 137, with only 0.08% asci with 2+ :6*c* in crosses, from which it differs in this property and in another respect. Crosses between alleles at locus *46* normally show no reciprocals (cross overs), but in crosses involving 277 about 11% of the 2+ :6*c* asci result from crossing over. The polarity of conversion criterion places 277 to the left of the 137 site.

The occurrence of the odd segregations (1+ :7*c*) reflect in mutant × mutant crosses the 5+ :3*c* observed sometimes in + × *c* crosses. The odd segregations appear to be the property of particular mutant genes or of the stocks in which they are carried. The genes 277 and 1216 at locus *46*, gene 60 at locus *19* and several at locus *75* have this property. It is notable that 277 does not produce 1+ :7*c* asci in crosses to 137, although it does so with all other locus *46* alleles.

The frequencies of conversion differ from one site to another in a locus. Often these are clearly polarized, decreasing from one end to the other. In some cases the differences in frequency of conversion are not in a systematic order (e.g. Rossignol 1969). Other differences found between different alleles concern the direction of conversion. In a cross + × *c*, four classes of exceptional asci are commonly observed in *Ascobolus immersus*, namely, 6+ :2*c*, 2+ :6*c*, 5+ :3*c* and 3+ :5*c*. The preferred direction of conversion is defined by the ratio of the frequencies of 6:2 to 2:6 or of 5:3 to 3:5 segregations, the wild-types being given first in each ratio. Differences of these kinds, comprising: (1) frequency of conversion or recombination; (2) the time of conversion whether meiotic (6:2 and 2:6) or post meiotic (5:3 and 3:5); and (3) direction of conversion, whether mutant to wild or the reverse, have been ascribed to the nature of the genetic alterations at the mutant sites and are called 'marker effects'. The time of segregation and the direction of conversion define a 'conversion spectrum'.

At the 164 locus, conversion is almost always meiotic (Baranowska, 1970). In crosses to wild-type, 6+ :2*w* and 2+ :6*w* frequencies are equal in the crosses 10 × +, 94 × + and 115 × +. They are unequal in other cases, conversion to + being commoner in W173 × +, 140 × + and 70 × +, while conversion to *w* is commoner in 24 × +. These differences bear no relation to the position of the mutant site in the locus. They are usually assumed to be due to the mutant itself rather than to any extraneous differences, but there is no proof in general of such 'marker' effects. The contrast with the parity rule shown in yeast is striking.

At locus *75*, the common exceptional asci in crosses of mutants to wild are also 2+ :6*c* and 6+ :2*c*. The ratio of these two classes (conversion to mutant divided by conversion to wild) constitutes a dissymetry coefficient (DC). The data depend upon counts of groups of eight ascospores, each group discharged from an ascus. A sample of the 2+ :6*c* and of the 6+ :2*c* was tested in each case for effects not due to conversion. Most spurious 2+ :6*c* asci were due to a tetratype segregation for a new (spontaneous) non-allelic mutant and the parent one. The few spurious 6+ :2*c* asci were due to two of the pigmented ascospores being genetically mutant. Rossignol (1969) found five distinct classes at locus *75*: α(DC = 11), β(1), γ(0.7), δ(0.04) and ε(2), so that the preferred directions of conversion were strongly different. The members of each DC class show a regular increase of conversion frequency from the left to the right end of the locus, the slope of increase ranging from a low value for α, through an intermediate one for γ to a high one for β (Fig. 2.5). Mutants belonging to different classes are mixed together in the locus. The observations could be accounted for by the formation of hybrid DNA from a fixed starting place and by different probabilities of correction according to the type of mispairing at a heterozygous site. An important observation concerns two deficiencies, w141 and 1303, constituting the δ type. In crosses to wild they show a great excess of 6+ :2*c*. This serves to identify one cause of bias in conversion. In hybrid DNA formed between a normal and deficient chain, a section of the mutant would leave a gap. Evidently a correction mechanism fills this with great efficiency. One would expect mutants due to an insertion to show the reverse effect, with a majority of 2+ :6*c* asci in crosses to wild type.

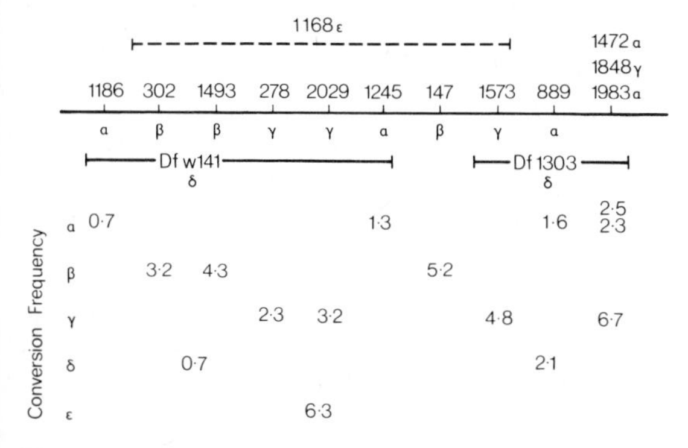

Fig. 2.5 *Ascobolus immersus* map of locus 75 showing the order of the sites of mutation in relation to their class of dissymmetry coefficient (α, β, γ, δ, ε) and the conversion frequency for each mutant site. The frequency of conversion increases from left to right for each class, but at different rates. Data of Rossignol (1969).

Correlation of these and other properties with molecular differences has been studied further by Leblon (1972) by examining the conversion spectra of mutants induced by particular mutagens. If the nature of the mutation were responsible for the observed differences in conversion spectra, mutants induced by mutagens having a very specific effect, producing only one kind of mutation, would have similar conversion spectra. However, mutants induced by mutagens having a low specificity should have a range of different conversion spectra. Different spontaneous mutants, the subject of most earlier work on *Ascobolus immersus*, exhibit all possible spectra.

Mutants at two loci (*b1* and *b2*), resulting in colourless spores, were induced in stock 28 by treating mycelium of one strain of one mating type with mutagen and crossing it to the other mating type. The use of just one strain is important since consistent differences between different specific mutagens would virtually exclude causation external to the locus itself. The mutants induced were 25 by an acridine (ICR170), 18 by a nitrosoguanidine (NG) and 19 by ethylmethane sulphonate (EMS). ICR170 induces frame shift mutants by insertion of an extra base pair or by deletion, with a small frequency of substitution of base pairs. The majority of NG mutations are thought to be base substitutions, transitions from AT to GC. On the other hand EMS is less specific. It acts mainly by alkylation of guanine causing transitions, preferentially of GC to AT, and frame shift mutations. All NG mutants showed frequent post-meiotic segregation and a predominance of conversion to wild type (Leblon's type C). In contrast, all ICR mutants showed no post meiotic segregation and a predominance of conversion to mutant (Leblon's type B). Most EMS mutants behaved like NG ones, but some showed only meiotic conversion with either predominance of conversion to wild type (type A) or to mutant (type B). Within these three broad classes there is quite a wide range of variation that is inexplicable in any very simple way (Table 2.13). There is also some indication of differences in pattern between

Table 2.13 *Ascobolus immersus*, conversion spectra of ascospore colour mutants induced by various mutagens at the *b1* and *b2* loci. The mutagens are an acridine (ICR), a nitrosoguanidine (NG) and ethylmethane sulphonate (EMS). The behaviour of the mutants in generating abnormal (conversion) asci in crosses to wild type form a basis for separating them into at least fifteen types. The ratios in columns 9 to 11 are respectively of meiotic to post meiotic (M/PM), of 6+ : 2*m*/2+ : 6*m*, and of 5+ : 3*m*/3+ : 5*m*. (Data of Leblon, 1972).

Mutagen	Locus	Mutant type	Frequencies of abnormal asci 6+ : 2*m*	2+ : 6*m*	5+ : 3*m*	3+ : 5*m*	Number of mutants	M/PM	6+/2+	5+/3+	Example
ICR	*b1*	1	12	1143	7	15	5	52	.01	.5	
	b2	1	185	2687	15	16	13	93	.07	1.0	A38
		2	157	207	1	0	2	364	.8	—	
		3	114	327	3	1	3	110	.4	—	
NG	*b1*	4	128	47	460	208	5	.26	2.7	2.2	
		5	36	11	30	24	1	.87	3.3	1.3	
		6	34	24	96	87	2	.32	1.4	1.1	
		7	5	0	29	2	1	.18	—	14	
		8	21	12	16	11	1	1.2	1.8	1.5	
	b2	8	111	75	109	58	2	1.1	1.5	1.9	
		9	55	66	21	24	1	2.7	.8	.9	
EMS	*b1*	10	155	78	174	173	5	.67	2	1	
		11	221	39	0	5	2	52	5.7	—	
		3	17	87	1	2	1	35	.2	—	
		12	7	17	34	18	1	.46	.4	1.9	
		13	22	2	28	15	1	.56	11	1.9	
		4	12	4	41	17	1	.28	3	2.4	
	b2	14	105	41	75	44	2	1.2	2.6	1.7	
		15	169	129	82	23	1	2.8	1.3	3.6	
		11	281	16	2	0	1	148	17.6	—	47E
		1	15	318	0	4	1	83	.05		

mutants at the *b1* and *b2* loci. There are perhaps no fewer than 15 kinds of conversion spectrum exhibited, some of these groups being heterogeneous. Tests to identify the molecular nature of the mutations, based on the probable mode of mutation, the response to mutagens in causing reversion and the occurrence and behaviour of intragenic suppressors, have led to the conclusion that one ICR mutant (*b2*A38) is a frame shift with an extra base and that one EMS mutant (*b2*47E) is a frame shift with a deletion. Both show little or no post meiotic segregation, with a preponderance of conversion to the mutant allele in the addition mutant or to the wild allele in the deletion. It is inferred that the single strand gaps expected to be formed in a heteroduplex are recognized efficiently and repaired by insertion of a base to give meiotic correction and segregation. It is not yet evident whether a unique nucleotide substitution, or a diversity, is characteristic of mutants which give post meiotic segregation. Theoretically, different mispaired bases might respond differently, though it is difficult to envisage so many different kinds of spectrum. One interesting observation was the discovery that one reversion of *b2*47E was accompanied by another genetic change that causes reduction in the frequency of conversion when heterozygous. It appears to be similar, if not identical, to the *cv* factors (see page 85).

Leblon and Rossignol (1973) have further analysed the behaviour of double mutants obtained by spontaneous reversion through intragenic suppression of an ICR (A38) and an EMS (A4) mutant. Ascospores of the double mutants are coloured light brown (A38, A2) or pink (A4, A5), contrasted with white in each single mutant and dark brown in the wild strain. A variety of aberrant types of ascus can be observed. One mutant (A38) and its suppressor (A2) both behave like frame shift mutants but of opposite type, respectively addition and deletion. The other mutant (A4) shows frequent post meiotic segregation on its own as though a base substitution, but its suppressor (A5) behaves as though it were an addition frame shift mutant. The range and frequencies of the fourteen principal types of ascus observed are given in Table 2.14. It will be seen that co-conversion is commoner by far than conversion of either single site suggesting that the sites are fairly close together. Conversion at single sites shows a preference for A4 over A5 and of A38 over A2, suggesting orders A4, A5 and A38, A2 respectively, with the first named of each pair nearer to the initiation site. These data, together with data from other crosses each segregating for a single site, are analysed for the spectrum of conversion at each mutant site in Table 2.15. It is evident that the pattern of conversion at A4 is affected by the presence of A5 in the same chromosome, while A38 and A2 affect one another mutually. The data as a whole are compatible with the view that conversions arise by correction in a heteroduplex and that these are often, if not usually, formed in both chromatids. The effect of the presence of A5, when heterozygous, is to cause frequent conversion at the A4 site and in the same direction. This proves that conversion by correction is induced at one mispaired site and

Table 2.14 *Ascobolus immersus*, conversion at the *b2* locus showing the types and frequencies of ascus found in crosses between wild and double site mutants, the two single site mutants in each of which act as partial mutual suppressors. Data of Leblon and Rossignol (1973). Phenotypes are numbers of white, coloured and dark spores per octad. Different ascus types of the same phenotype were separated by tests of progeny. Fractional numbers are due to scaling down of initial numbers when some of a sample were not confirmed by tests.

Ascus types							
	1 2	1 2	1 2	1 2	1 2	1 2	1 2
	1 2	1 2	1 2	1 2	1 2	1 2	1 2
	1 2	1 2	+ 2	1 2	1 2	1 2	1 2
	1 2	1 2	+ 2	1 2	+ 2	+ 2	1 2
	+ +	1 +	+ +	1 +	+ +	1 +	1 2
	+ +	1 +	+ +	+ +	+ +	+ +	1 2
	+ +	+ +	+ +	+ +	+ +	+ +	+ +
	+ +	+ +	+ +	+ +	+ +	+ +	+ +
Phenotype	0:4:4	2:4:2	2:2:4	1:4:3	1:3:4	2:3:3	0:6:2
A5 A4 / + +	25765	4	6.7	96	70	19	1896
A38 A2 / + +	9997	2.8	5.6	0	0	0	377

Ascus types							
	1 2	1 2	1 2	1 2	1 2	1 2	1 2
	1 2	1 2	1 2	1 2	1 2	1 2	1 2
	+ +	1 2	1 2	1 2	1 +	1 +	+ 2
	+ +	1 2	+ 2	1 2	1 +	+ +	+ 2
	+ +	+ 2	1 2	1 2	+ +	+ +	+ 2
	+ +	+ 2	+ 2	+ 2	+ +	+ +	+ 2
	+ +	+ +	+ +	+ +	+ +	+ +	+ +
	+ +	+ +	+ +	+ +	+ +	+ +	+ +
Phenotype	0:2:6	2:4:2	2:4:2	1:5:2	2:2:4	1:2:5	4:2:2
A5 A4 / + +	133	22	8	275	1.3	11	0
A38 A2 / + +	481	2.8	0	0	0.9	0	1.7

that the effect often travels along to a second mispaired site on the same chain of the same chromatid. A4 has no detectable effect on A5 because it is usually not corrected at the meiotic division and postponement of correction at one site can have no influence on another site that is not prone to post meiotic segregation. The A38 A2 case shows that either site may induce correction and that the direction of conversion whether to mutant or to wild is determined by the preference exercised at the first site corrected. From the first site the effect travels along the same DNA chain to the second site. Moreover, since each site affects the other about equally, there is no polarity in the process of conversion, even though there is polarity in the establishment of hybrid DNA in the chromatids.

Table 2.15 *Ascobolus immersus* conversion at the *b2* locus, illustrating the influence of mutation at one site in a locus on conversion at a second site in the same locus. The conversion segregations are expressed as numbers of asci per thousand, the rate of conversion at the *b2* locus being 8 to 17‰. Data of Leblon and Rossignol, 1973.

	1+ × ++	12 × +2	12 × ++ Seg. 1/+	12 × ++ Seg. 2/+	12 × 1+	+2 × ++
A4, A5						
6+ : 2*m*	7	1	5	5	3	5
2+ : 6*m*	9	6	67	78	83	78
5+ : 3*m*	45	28	3	0	0	0
3+ : 5*m*	46	35	13	0	0	0
ab4+ : 4*m*	—	—	1	—	—	—
A38, A2						
6+ : 2*m*	12	8	44	45	112	142
2+ : 6*m*	154	99	35	35	11	12
5+ : 3*m*	1.5	0	0	0	1	1
3+ : 5*m*	0.5	0	0	0	0	0

The relation of conversion to segregation of neighbouring markers has been examined by Stadler, Towe and Rossignol (1970) with respect to the *w17* locus, using the Pasadena strains P5(−) and K5(+). The *w17* locus is flanked to the left by *colt* (colonial at a restrictive temperature) and to the right by *fpr* (conferring resistance to fluorophenylalanine). The spacing is *colt* (10) *w17*(3) *fpr*. Conversion of a particular mutant is subject to genetic influences extrinsic to the *w17* locus; at least two mutants show significant fluctuations of at least four fold differences in frequency. The interpretation of recombination frequencies as well as the ordering of the sites is limited by these effects, though the data are compatible with an order based on increasing frequencies of conversion, the highest values being to the right. In a sample of 2+ : 6*w* asci from allelic crosses there were, in respect of the flankers, 228 PD : 12 NPD : 194 T. That is, as in yeast, crossing over in the neighbourhood of a conversion occurs as frequently as not.

2.5 Sordaria fimicola

The genus *Sordaria* is similar in many respects to *Neurospora*. However, *S. fimicola* is homothallic, so that sexual reproduction and the formation of perithecia and asci follows the association of two nuclei that may be completely alike. Nearly all work with it has used colour mutants of ascospores at the *gray* (*g*) locus. Different alleles are gray to colourless (often referred to as hyaline, *h*) compared with the darkly pigmented normal. All mutants commonly used are derived from one wild strain

(A1), the source of the original *g* mutant, which is only poorly cross fertile with strains other than A1.

Crosses of *g*×+ show principally six classes of asci, namely regular 4+:4*g* and five kinds of conversions, respectively 6+:2*g*, 2+:6*g*, 5+:3*g*, 3+:5*g* and aberrant 4+:4*g*. The aberrant 4+:4*g* are separated from truly normal 4+:4*g* that mimic them (through spore displacement) by following linked or unlinked factors, such as the ascospore colour mutant *indigo*. The frequency of conversion is about one per 430 asci for all alleles examined. The spectrum differs from one allele to another (Table 2.16). Most show a bias towards post meiotic segregation and a range from bias of *g*→+ conversion to +→*g* conversion.

Table 2.16 *Sordaria fimicola*, effect of one allele on segregation of a second at the *gray* locus. Data of Kitani and Olive (1967; 1969); *classification only for segregation of *h* and +.

	Abnormal segregations				
Cross	6+:2*g*	2+:6*g*	5+:3*g*	3+:5*g*	ab4+:4*g*
g1×+	31	6	35	14	13
h2×+	7	5	26	23	35
**h2*+×+*g*	6	20	25	31	17
h2a×+	6	4	28	12	47
**h2a*+×+*g*	11	14	28	31	16
h3×+	3	13	9	40	35
**h3*+×+*g*	2	11	27	40	19
h3a×+	0	0	21	9	19
h4×+	1	4	4	28	16
**h4*+×+*g*	4	8	17	21	12
**h4g*×++	6	5	12	10	8
h4b×+	2	3	11	20	11
**h4b*+×+*g*	2	9	21	35	13

If closely linked factors either side of *g* (Fig. 2.6), are followed, fourteen classes of asci with conversions are readily distinguishable. The data (Table 2.17) show that, except for the 6+:2*g* class, recombination of outside markers occurs in half of the converted asci (Kitani and Olive, 1967). Analyses of crosses between alleles at the *gray* locus have been reported (Kitani and Olive, 1969) in respect of a series of *g*×various *h* genes and *g h4*×+. A large range of abnormal asci was observed. These can be classified for segregation at the *h* site, but not at the *g* site without further tests, since the hyaline *h* mutation is epistatic to the *g* mutation. The data included in Table 2.16 do not show any striking effects of *g* on any *h*.

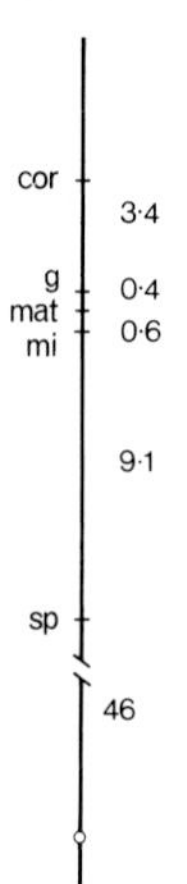

Fig. 2.6 *Sordaria fimicola* map of linkage group 1 showing relative location of *gray* ascospores (*g*) gene to mycelial genes *spotty* (*sp*), *milky* (*mi*), *mat* and *corona* (*cor*).

Table 2.17 *Sordaria fimicola*, recombination data for abnormal asci from crosses of the type *mat*++ × +*g cor*. The ascus types are those listed in Table 2.18, with *mat* ≡ *P*, *g* ≡ *m* and *cor* ≡ *d*. The odd numbered types of ascus have the parental combinations of *mat*+ and +*cor*, while the even numbered types have *mat cor* and reciprocal ++ recombinants. (Data of Kitani and Olive, 1967).

Segregation at *g* locus	6+:2*g*		2+:6*g*		ab4+:4*g*		5+:3*g*				3+:5*g*			
Ascus type	3	4	1	2	9	10	17	15	16	18	13	11	12	14
g allele														
g	55	22	4	6	6	4	29	21	25	13	13	2	4	3
h2	4	1	4	0	17	13	13	3	3	8	12	7	2	10
h3	1	0	6	5	21	29	9	6	1	5	37	4	8	13
h4	1	0	5	0	10	11	3	0	3	3	16	5	7	10

2.6 Symmetrical or asymmetrical hybrid duplexes?

In principle, the types of abnormal tetrad observed can be explained by the occurrence of heteroduplexes in non-sister chromatids and by the 'correction' of mismatched bases at sites of difference between the chromosomes. A significant question to attempt to answer is whether heteroduplexes are formed symmetrically, to equal extents in the two

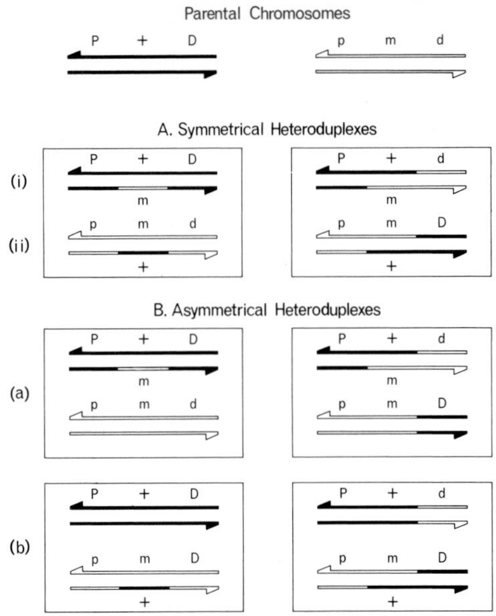

Parental Chromosomes

P + D　　p m d

A. Symmetrical Heteroduplexes

(i) P + D / m　　P + d / m

(ii) p m d / +　　p m D / +

B. Asymmetrical Heteroduplexes

(a) P + D / m / p m d　　P + d / m / p m D

(b) P + D / p m D / +　　P + d / p m D / +

Fig. 2.7 Source of types of conversion asci, listed in Table 2.18, from symmetrical heteroduplexes (A) and asymmetrical heteroduplexes (B). Each box contains a diagram of two of the four chromatids. In each case there will also be two more chromatids, one of each parental type. The flanking regions are parental in the first column of A and B and recombined in the second column. Corrections ($m \rightarrow +$ or $+ \rightarrow m$) in the heteroduplex segments lead to the ascus types listed in Table 2.18, as follows:

Corrections		Flanking markers	
		Parental	Recombined
in A(i) and A(ii)			
$+ \rightarrow m$	$+ \rightarrow m$	1	2
$m \rightarrow +$	$m \rightarrow +$	3	4
$+ \rightarrow m$	$m \rightarrow +$	5	6
$m \rightarrow +$	$+ \rightarrow m$	7	8
$+ \rightarrow m$	none	11	12
$m \rightarrow +$	none	17	18
none	$+ \rightarrow m$	13	14
none	$m \rightarrow +$	15	16
none	none	9	10
in B(a)			
	$+ \rightarrow m$	1	2
	m $\rightarrow +$	7	8
	none	13	14
in B(b)			
	$+ \rightarrow m$	7	8
	$m \rightarrow +$	3	4
	none	17	18

Table 2.18 Possible types of ordered octad asci arising from the cross $P+D \times pmd$, assuming correction in symmetrical (s) or asymmetrical (a) heteroduplexes covering the m locus (see Fig. 2.7). Types 1 to 8 are detectable in unordered asci, such as the tetrads of yeast.

Ascus type	1	3	5	7	9	11	13	15	17
	$P+D$	$P+D$	$P+D$	$P+D$	$P+D$	$P+D$	$P+D$	$P+D$	$P+D$
	$P+D$	$P+D$	$P+D$	$P+D$	$P+D$	$P+D$	$P+D$	$P+D$	$P+D$
	PmD	$P+D$	PmD	$P+D$	$P+D$	PmD	$P+D$	$P+D$	$P+D$
	PmD	$P+D$	PmD	$P+D$	PmD	PmD	PmD	PmD	$P+D$
	pmd	$p+d$	$p+d$	pmd	$p+d$	$p+d$	pmd	$p+d$	$p+d$
	pmd	$p+d$	$p+d$	pmd	pmd	pmd	pmd	$p+d$	pmd
	pmd	pmd	pmd	pmd	pmd	pmd	pmd	pmd	pmd
	pmd	pmd	pmd	pmd	pmd	pmd	pmd	pmd	pmd
Occurrence	s	s	s	s	s	s	s	s	s
	a	a		a(2)			a		a
Ascus type	**2**	**4**	**6**	**8**	**10**	**12**	**14**	**16**	**18**
	$P+D$	$P+D$	$P+D$	$P+D$	$P+D$	$P+D$	$P+D$	$P+D$	$P+D$
	$P+D$	$P+D$	$P+D$	$P+D$	$P+D$	$P+D$	$P+D$	$P+D$	$P+D$
	Pmd	$P+d$	Pmd	$P+d$	$P+d$	Pmd	$P+d$	$P+d$	$P+d$
	Pmd	$P+d$	Pmd	$P+d$	Pmd	Pmd	Pmd	Pmd	$P+d$
	pmD	$p+D$	$p+D$	pmD	$p+D$	$p+D$	pmD	$p+D$	$p+D$
	pmD	$p+D$	$p+D$	pmD	pmD	pmD	pmD	$p+D$	pmD
	pmd	pmd	pmd	pmd	pmd	pmd	pmd	pmd	pmd
	pmd	pmd	pmd	pmd	pmd	pmd	pmd	pmd	pmd
Occurrence	s	s	s	s	s	s	s	s	s
	a	a		a(2)			a		a

chromatids, or asymmetrically, to unequal extents ranging from none in one chromatid to less in one chromatid than in the other. The effects of the extremes on a point of hybrid difference, flanked closely each side by genetic markers (Fig. 2.7), may be considered. The flanking markers may be recombined as part of the event around a heteroduplex or not, with equal frequency. A point of hybridity ($+/m$) may be corrected to normal ($+$) or to mutant (m) during the first division of meiosis or be left uncorrected until the first post meiotic mitosis. Each of these modes of correction may occur independently in each heteroduplex. Hence, in the symmetrical case, there can be 18 types of ascus (Table 2.18) consequent, taking into account the flanking markers. However, in the asymmetrical case, there can be only ten types, each like one of the 18 arising from the symmetrical heteroduplexes. The occurrence of the remaining eight types and their frequencies relative to the ten in common will indicate the probability with which the symmetrical and asymmetrical states occur. Critical evidence is scarce, the question having been considered critically in only two cases. Both of these indicate a predominance of asymmetry.

The eighteen possible types distinguishable in an octad, arising from heteroduplex symmetry, are listed in Table 2.18, those with flankers in parental combination having odd numbers and those with flankers recombined around the heteroduplexes having even numbers. The patterns shown group together permutations equivalent to the ones shown. In an organism in which correction occurs regularly during meiosis only types 1 to 8 are expected. In respect of the site showing conversion there are three classes: 2+ : 6*m*(1,2), 6+ : 2*m*(3,4) and 4+ : 4*m*(5–8). In an organism having tetrads of spores, like yeast, rather than octads, the three classes would be: 1+ : 3*m*, 3+ : 1*m* and 2+ : 2*m*. The last class appears to be normal and in respect of the flanking markers there are four subclasses: parentals (7), two strand doubles (5), recombinants between the *p* and *m* loci (6) and recombinants between the *m* and *d* loci (8). If each of the modes of correction is equally frequent, each of these subclasses should arise from a heteroduplex with a quarter of the total frequency of the 1:3 and 3:1 conversions. The assumption of equality of probability of the events is fundamentally reasonable; if +→*m* and *m*→+ corrections were unequal, the calculations are more complex but the argument not very different. If heteroduplexes were formed asymmetrically, in only one chromatid, type 5 among those with meiotic correction could not occur except as the result of a two strand double cross over, one cross over in each of the short intervals, and so likely to be rare. In yeast, Fogel and Mortimer (1974) report 319 *arg4* conversions (ascus types 1 to 4) among 3734 unselected tetrads from diploids of the type P+D/p*arg4*d; 148 of the conversions were recombined for the flanking markers (ascus types 2 and 4). If conversion at *arg4* were contingent upon independent repair in two heteroduplexes about 80 apparent two strand doubles would be expected. In fact, only 14 two strand doubles (ascus type 5) were observed, together with 9 three strand and 6 four strand doubles. The facts favour a single heteroduplex or a restraint on two heteroduplexes such that correction must usually involve DNA strands of opposite polarity in the two hybrid regions. The properties of co-conversion require all correction in each instance to be in the same strand.

Stadler and Towe (1971) have obtained similar indications for *Ascobolus immersus* utilizing 5+ : 3*m* and 3+ : 5*m* asci, segregating at the *w17* locus and marked to the left with *colt* and to the right with *fpr* (see page 39). They distinguish between four types of ascus, two with flanking markers in parental conformation (types ST1 and ST2 corresponding respectively to types 11 and 15 and types 13 and 17 of Table 2.18, neglecting the direction of conversion) and two (types ST3 and ST4) with flanking markers recombined to produce tetratype asci. Type ST3, which corresponds to types 14 and 18 of Table 2.18, neglecting the direction of conversion, can arise from either two hybrid chromatids or from one, by a cross over involving the chromatid in which there is a mispaired base not corrected until after meiosis. Type ST4 does not correspond to any type in Table 2.18; it has a cross over between the flankers not involving the chromatid with a hybrid section (and conversion). Stadler and Towe state that only the one hybrid state could yield such a type. However, it could arise from a two hybrid state with a cross over between a pair of chromatids that does not include the one undergoing post meiotic segregation. They do not identify a class corresponding to types 12 and 16 of Table 2.18; these could arise only from symmetrical hybridity. In Table 2.19 the data show that, for types ST1 and ST2, there is strong asymmetry in formation of heteroduplexes. Kitani and Olive's (1967) analysis of 5:3 and 3:5 segregations at the *g* locus in *Sordaria fimicola* also indicates a difference, though of lesser degree. Their 'substitution' classes correspond to ascus types 11, 12, 15 and 16 and their 'restoration' classes to ascus types 13, 14, 17 and 18.

Table 2.19 Data for 5+:3*m* and 3+:5*m* asci, also segregating for flanking markers, showing asymmetry in formation of heteroduplexes. Data: *Ascobolus*, Stadler and Towe, 1971; *Sordaria*, Kitani and Olive, 1967.

Segregation	5+:3*m*				3+:5*m*			
Ascus type (Table 2.18)	15	17	18	16	11	13	14	12
expected symmetry	$\frac{1}{4}$	$\frac{1}{4}$	$\frac{1}{4}$	$\frac{1}{4}$	$\frac{1}{4}$	$\frac{1}{4}$	$\frac{1}{4}$	$\frac{1}{4}$
expected asymmetry	0	$\frac{1}{2}$	$\frac{1}{2}$	0	0	$\frac{1}{2}$	$\frac{1}{2}$	0
Ascobolus immersus	0	21	—	—	1	32	—	—
Sordaria fimicola	33	71	42	41	11	91	24	46

It may be noted that aberrant 4+:4*m* segregations (ascus types 9 and 10) are not expected from asymmetric heteroduplexes; however they have been reported to be common in *Sordaria*.

Stadler and Towe further applied the analysis to recombination between alleles at the *w17* locus, using data from crosses in which *w17a* or *w17b* (both near one end of the locus and showing about 0.41% of 6+:2*m* asci in crosses to wild) were mated to other alleles towards the other end of the locus (average 0.024% of 6+:2*m* asci in crosses to wild). The treatment involves assumptions for values of the ratio of one chromatid to two chromatid events, the conversion frequencies of the two alleles and the length of the hybrid DNA relative to the distance separating the two alleles. They also assume that one segregating site is in the hybrid DNA and the other not. These considerations would lead to equal numbers of reciprocal and non reciprocal recombinants between the two *w17* alleles arising from events with two hybrid chromatids. In fact analyses show that reciprocal recombination occurs in no more than about an eighth of the asci; more exactly 25 per 195 asci. Hence the proportion of 2+:6*w* asci arising from meiosis with two hybrid chromatids would be twice this value or 0.26. Other estimates are for the relative frequencies of non-reciprocal events at the two sites: 0.88 at *w17a* or *w17b* and 0.12 at *w17e*. An alternative, based on frequencies of conversion in crosses to wild, leads to 0.94 and 0.06 respectively. In a sample of 383 asci the agreement of the fit to expectations from a heteroduplex in one chromatid is very good (Table 2.20).

Table 2.20 *Ascobolus immersus*, observed 2+:6*m* asci in allelic crosses and expectations if a heteroduplex is formed on one chromatid only. Data of Stadler and Towe (1971). All crosses are of form $Pm^1+D \times p+m^2d$, with mutant site 1 proximal to site 2. In the classification for the flanking genes, P.D. = parental ditype, T. = tetratype and N.P.D. = non parental ditype.

m^+ spore in	*PD*		*pd*		*pD*		*Pd*	
Flanking genes	P.D.	T.	P.D.	T.	T.	N.P.D.	T.	N.P.D.
Observed	191	34	15	7	107	6	23	0
Expected:								
(a) 0.88:0.12	185	34	25	4	111	0	26	0
(b) 0.94:0.06	197	34	13	2	109	0	27	0

2.7 Map Expansion

A curious property of fine structure maps in fungi is map expansion. The recombination observed between two relatively distant sites of difference in a locus tends to be greater than the sum of the values observed over a number of short intervening distances. Explanation of this effect is elusive. Fincham and Holliday (1970) argue that the chance of a mismatched base pair, in a segment of heteroduplex DNA, being corrected independently of another linked mismatched pair will increase sharply as the distance between the two sites becomes greater than the length of the DNA segment involved in the correction process. It is probable that correction of mismatches succeeds establishment of heteroduplexes. Correction could start in the neighbourhood of any mismatch and events starting in different places, more likely if relatively remote, need not be correlated in their attack.

If the probability of independent correction increased with separation, the mapping curve should show three phases: (1) an initial additive section where the recombining sites are close; (2) a section of increased slope where the sites are further apart and map expansion is found; (3) a final additive section of reduced slope beyond the region manifesting expansion. A comparison of the initial and final slopes should give information about the relation between gene conversion and crossing over. While many sets of experimental data show the transition from (1) to (2), very few show the second transition. The only cases which are fairly convincing are those for the *paba-1* locus in *Aspergillus* (Siddiqui, 1962) and the *46* locus in *Ascobolus* (Rossignol, 1964). Using data from genes which control known products, estimates of the minimum length of the DNA segments involved in correction are about 40 nucleotides in fission yeast and at least 130 nucleotides in *Neurospora*. Ahmad and Leupold (1973) approach an explanation differently, attempting to correlate map expansion with reciprocal recombination based on crossing over.

2.8 Drosophila melanogaster

The notion of the indissolubility of the gene by recombination in a heterozygote died slowly in Drosophilology. In cases in which there was physiological evidence of allelism, $m^1+/+m^2$ being mutant in phenotype while $m^1m^2/++$ was wild type, and m^1 and m^2 could be recombined from a heterozygote they were regarded as pseudoalleles, falsely allelic as though the physiological test was in error. However, cases accumulated in which several similarly acting mutants had to be regarded as pseudoalleles. Hence abandonment of the notions of indissolubility and of pseudoallelism is necessary.

In an early study, Chovnick (1961) examined eight of the ten possible heterozygotes between five alleles at the *garnet* locus (linkage group 1-44.4). Seven of these crosses yielded g^+ offspring, at the rate of about 2 or 3 per 10^5 progeny. All crosses were marked at positions close (1.4 to 8.2 map units) to each side of the *garnet* locus. Collectively, in the crosses of general type $a\,g^x\,b \times A\,g^y\,B$ the g^+ progeny observed were: 4 *a b*, 3 *A B*, 11 *a B* and 5 *A b*. Thus, although a majority of the g^+ progeny were accompanied by recombination of flanking genes, nearly a third were not. Moreover, both types of recombinant progeny were present in two cases. These observations are consistent with a fair proportion of the g^+ progeny arising by conversion. Evidence has continued to accumulate favouring the conclusion that recombination in the fly is essentially like that in fungi. However, the analysis of fine structure of loci in *Drosophila melanogaster* is relatively primitive due to the low frequency and to the labour of scanning large numbers of progeny.

By far the best data derive from studies of the *maroon-like* (*mal*, 1-64.8) and *rosy* (*ry*, 3-52.0) loci. Mutants at these loci are deficient for xanthine dehydrogenase, for which ry^+ is the structural gene and *mal* a regulator, and mutant larvae die if they are fed on medium that contains an excess of purine. However, wild type strains live in the presence of excess purine so the rare prototrophic recombinants from crosses between different *mal* (or *ry*) mutants may be selected. Since the locus of *mal* is in the X chromosome, half tetrads can be recovered using attached-X stocks. In the case of *ry*, compound-3 chromosomes can be used to recover half tetrads. In these compound-3 strains, the two normal chromosome 3s are replaced by two chromosomes consisting respectively of two left arms, C(3L), and two right arms, C(3R). Determination of the constitution of each recombinant mal^+ or ry^+ half tetrad individual requires detachment of the arms and their separate analysis. Prototrophs are formed about ten times more frequently at the *ry* than at the *mal* locus and mutants occur more frequently at the *ry* locus (Fig. 2.8). Much more extensive data, confirming the conclusions from the *garnet* studies, are now available (Chovnick *et al.*, 1969, 1970, 1971; Finnerty *et al.*, 1970; Smith *et al.*, 1970).

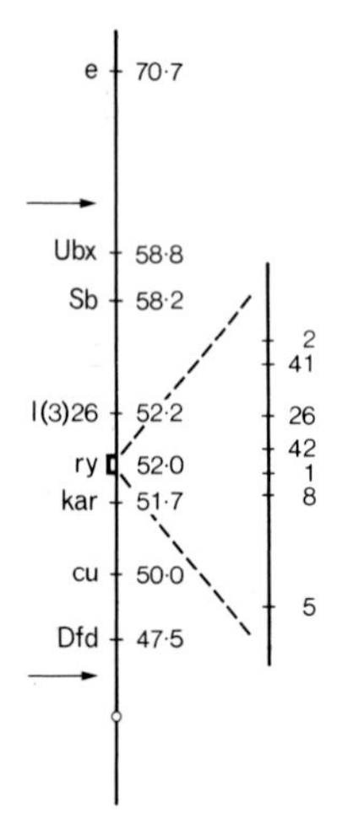

Fig. 2.8 *Drosophila melanogaster*, map of *rosy* (*ry*) locus and its environs, together with a fine structure map of mutant sites in *rosy*. The labelled gene loci are: *Dfd*, *Deformed*; *cu*, *curled*; *kar*, *karmoisin*; *ry*, *rosy*; *l(3)26*, a recessive lethal; *Sb*, *Stubble*; *Ubx*, *Ultrabithorax*; *e*, *ebony*. The arrows mark the limits of the inversion *In(3R)P18*.

Data are summarized in Table 2.21. In crosses of the type $A\,ry^x\,B \times a\,ry^y\,b$, three classes of progeny are observed among ry^+ individuals, namely the recombinant *A b* (or *a B*) and the two parentals *A B* and *a b*. The first may be regarded as the result of a classical crossing over (serving to order the *ry* sites), the second as conversion of ry^x to ry^+ and the third as conversion of ry^y to ry^+. The proportion of cross overs rises as the genetic distance between the sites of difference of the *ry* mutants increases. Among the converts, the right hand (more distal) *ry* site shows more frequent conversion, except that ry^5 predominates in several combinations (with ry^{42}, ry^1 and ry^2); however, ry^2 gives peculiarly low values in all crosses and may be a single site deletion or a small inversion. Thus conversion in the *rosy* locus appears polarized with frequencies decreasing from right to left. Analysis of half tetrads shows that some at least of those with a ry^+ and a reciprocal exchange of the flanking markers in fact arose from a non-reciprocal event of conversion associated with exchange of the flankers. The constitutions of the parent and a critical ry^+ progeny, with a half tetrad, were:

$$\frac{\text{Dfd} + kar^{3l}\ \text{ry}^2\ +\ +}{+\ \text{cu kar}\ \ \text{ry}^{14}\ 126\ \text{Sb}} \rightarrow \frac{\text{Dfd} + kar^{3l}\ +\ 126\ \text{Sb}}{+\ \text{cu kar}\ \ \text{ry}^2\ +\ +}$$

The effect of a small deletion (*kar*^{3l}) to the left of the *rosy* locus is to reduce recombination, particularly of the cross over class and of conversion at the left hand *ry* site, with which it entered the cross. It appears that differences in background genotypes may affect both crossing over and conversion at the *rosy* locus. In Table 2.21, the *ry*^{41} × *ry*^{5} crosses in lines 6 and 11 were constructed from different stocks and that in line 11 shows significantly higher frequencies of crossing over and of conversion of *ry*^{41}. A multiple rearrangement in chromosome 2 has the effect of increasing recombination at the *ry* locus, both by crossing over and by conversion. When one of the *ry* alleles is in an inversion, the frequency of the *ry*$^+$ with a cross over of the flanking genes is dramatically reduced, but the frequencies of converts with parental combinations of flankers are not much reduced (Table 2.21b). The data were derived (Chovnick, 1973)

Table 2.21 *Drosophila melanogaster*, crossing over and conversion at the *rosy* locus. (a) All crosses of the type *kar ry*^{x} *l(3)26* × +*ry*^{y}+; (b) crosses of the type *kar ry*^{41} *l(3)26* × *InP18*+*ry*^{y}+. Data of (a) Chovnick, Ballantyne and Holm, 1971, (b) Chovnick, 1973.

Alleles in cross *ry*^{x}	*ry*^{y}	*ry*$^+$ per 10^6	Classification of *ry*$^+$ cross overs	conversion *ry*^{x}	conversion *ry*^{y}
(a) *41*	*2*	3.6	0	8	1
41	*26*	22.5	3	10	3
41	*42*	22.8	5	18	7
41	*1*	28.1	11	10	2
41	*8*	32.4	11	18	11
41	*5*	41.3	21	20	11
5	*8*	12.0	0	2	7
5	*1*	14.6	3	5	2
5	*42*	25.8	10	7	3
5	*2*	24.3	28	10	1
5	*41*	58.2	40	8	30
2	*42*	6.9	0	0	4
2	*1*	27.1	20	0	6
26	*42*	20.1	1	8	5
42	*8*	11.9	2	4	3
42	*5*	21.8	5	4	5
(b) *41*	*26*	8.6	0	4	2
41	*42*	15.4	0	15	5
41	*8*	14.4	0	9	4
41	*5*	13.3	1	21	6

from inversion heterozygotes involving *In(3R)P18*, a relatively short inversion with the *rosy* locus at about its centre (Fig. 2.8). Evidently, there is plenty of transfer of information between inversions in a heterozygote, except perhaps close to the break points.

The conclusion from work with *Drosophila* is that the basic event in recombination involves a non-reciprocal transfer of information, conversion. It may or may not be accompanied by a reciprocal event in its vicinity. Like crossing over, conversion is observed in mutant heterozygotes but not in homozygotes. It occurs in females, but not in males. The wild type alleles generated by conversion in heterozygotes of various allelic mutants are identical to one another, to wild types that are apparently cross overs and to the wild type in the control stock. Conversion, like crossing over, may be suppressed by heterozygous structural rearrangements with breaks close to the region concerned.

2.9 Zea mays

Evidence that recombination between alleles at the *waxy* (*wx*) locus is at least in part due to conversion is provided by observations of Nelson (1962, 1968). Mutation at the *waxy* locus results in an absence of amylose from the pollen grains, embryo sacs and the endosperm of the kernel. In consequence, the 'starch' grains stain reddish brown with iodine solution rather than deep blue. The kernels are relatively dull in appearance. By introducing another gene *ae* (*amylose extender*) into the stock, the kernels have a somewhat crinkled or shrivelled appearance when dry, like a *sugary* phenotype. In hybrids between different *waxy* alleles a few wx^{+} pollen grains are found in the anthers and, if back crosses to a homozygous *wx* stock are made, a few wx^{+} kernels are found. In the cross $wx^{90} \times wx^{Coe}$, recombinant kernels were found with a frequency of 99 per 10^5, whereas recombinant pollen grains were 102 per 10^5. These frequencies are the same. In crosses in which gene markers are used each side of the *wx* locus in the cross, all combinations of these markers appear among the wx^{+} recombinants (Table 2.22). One recombinant class is predominant, indicating the order of mutant sites at the *wx* locus with

Table 2.22 *Zea mays*, segregation of flanking markers in allelic recombinants. All crosses were of the type $Pm^1D \times pm^2d$. In the *wx* crosses P/p were $+/v1$ and D/d were $+/bz$. In the *gl1* crosses P/p were $+/o2$ and D/d were $+/sl$.

Parental alleles		m^{+} progeny				
m^1	m^2	*PD*	*pd*	*pD*	*Pd*	Source of data
wx^{90}	wx^{Coe}	16	36	81	4	Nelson, 1962, 1968
wx^{Coe}	wx^{H21}	5	0	3	0	Nelson, 1968
$gl1^{x}$	$gl1^{st}$	19	21	19	12	Salamini and Lorenzoni, 1970

respect to the flanking loci; this is *bz* H21 Coe 90 *v*, the *v* (*virescent*) locus being proximal.

Salamini and Lorenzoni (1970) have found high frequencies of recombination between alleles at the *gl1* locus, ranging from 0.06% in *gl1*st × *gl1*^{i} to 0.51% in *gl1*st × *gl1*^{h}. Using outside markers *o2* and *sl*, mapped as *o2* (20) *gl1* (14) *sl*, they found high numbers of recombinants with parental combinations of outside markers (Table 2.22), comparing the data to the 'high negative interference' effect in microorganisms.

3

Mitotic recombination

And where it cometh, all things are;
And it cometh everywhere.
R. W. Emerson, *The Absorbing Soul*

The occasional occurrence of crossing over between homologous chromosomes at mitosis has been observed. Stern (1936) reported extensively on the phenomenon in *Drosophila melanogaster*, in respect of flies heterozygous for genes whose homozygous effect can be recognized in a small spot. Using the linked genes *yellow* body colour (y) and *singed* bristles (sn), Stern found that heterozygous females of the constitution $\frac{y\ +}{+\ sn}$, phenotypically not yellow nor singed, bore occasional spots that were *yellow* and others that were *singed* and that these frequently appeared contiguously, as twin spots (Table 3.1). The frequency of these events is accentuated by other genetic factors, especially *Minute* factors. On the other hand, in females of the constitution $\frac{y\ sn}{+\ +}$, there were no twin spots and the single spots were mostly $y\,sn$, less commonly y and very rarely sn (Table 3.1).

Table 3.1 Spots observed in *Drosophila melanogaster* as a consequence of somatic segregation.

Constitution	y	sn	twin y, sn	$y\,sn$
$\frac{y\ +}{+\ sn}$	124	234	147	0
$\frac{y\ sn}{+\ +}$	43	7	0	110

Presumably the $y\,sn$ spots were often accompanied by a $\frac{+\ +}{+\ +}$ spot, not detectable phenotypically.

Stern explained the results as a consequence of crossing over at mitosis between two out of four chromatids (Fig. 3.1), precisely equivalent to classical reciprocal recombination at meiosis. Crossing over may occur between the centromere and *sn* or between *sn* and *y*. The two sister chromatids of each chromosome separate to opposite poles at anaphase and the two homologous chromosomes orient independently. Hence the recombinant chromatids may go to the same pole and the event be not detectable, or they may go to opposite poles and be recognized by the phenotypic consequences as single or twin spots. Fig. 3.1 illustrates the orientations that yield homozygous cells and shows that, for $\frac{y\ +}{+\ sn}$, twin spots result from recombination in the centromere to *sn* interval, while *y* spots result from recombination in the *sn* to *y* interval. The occurrence of 'twin' spots shows that the phenomenon is frequently reciprocal. On the theory outlined, the solitary *sn* spots require double crossing over. However, their very high frequency suggests that somatic recombination is not always reciprocal. Further analysis is prevented because the recombinant tissue cannot be isolated and propagated. Significant developments came from the discovery of the same phenomenon in fungi.

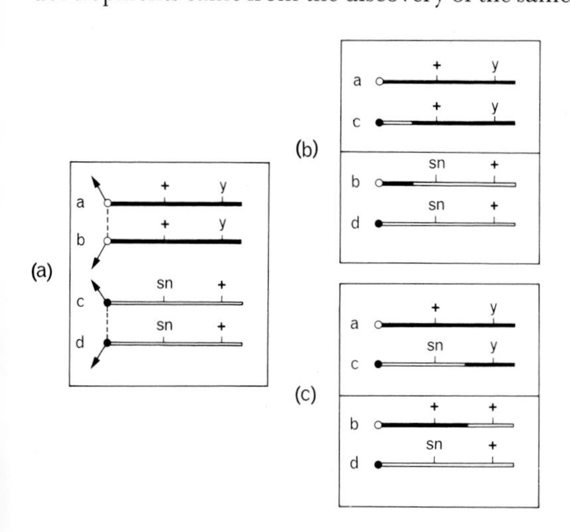

Fig. 3.1 Diagram to illustrate the genesis of single and twin spots in *Drosophila melanogaster* by mitotic crossing over. In the parental cell (a) arrows indicate the direction of the mitotic pole to which chromatids segregate, chromatids a and c ending in one cell, b and d in the other. After a cross over between centromere and *sn* (*singed* bristles) a twin 'mutant' spot results (b). After a cross over between *sn* and *y* (*yellow* body colour) a single 'mutant' spot results (c).

Methods were found of inducing diploidy in habitually haploid filamentous fungi, first of all by Roper (1952, 1955) with *Aspergillus nidulans* by selecting wild type progeny from heterocaryons, using either two conidial colour mutants (*white* and *yellow*) or two auxotrophs. The method using auxotrophs has been applied successfully to a number of other moulds with uninucleate conidia, but has failed with *Neurospora crassa*. In the latter it seems impossible to maintain strains with diploid nuclei or nuclei with extra chromosomes. There is rapid disjunction of the homologues, as in pseudo-wild types (Mitchell, *et al*., 1952), leading to more or less complex heterocaryons with only haploid nuclei. In yeast, as already noted, diploidy is easy to establish and maintain, so long as entry into meiosis is not induced.

3.1 Aspergillus nidulans

Diploid strains of normally haploid fungi tend to produce sectors showing segregation of alleles present originally in heterozygous condition. The nuclei of the segregant cells are sometimes haploid and sometimes still diploid but homozygous with respect to genes originally heterozygous. Analysis depends upon picking segregants visually for colour or by selective isolation for nutritional requirements or resistance to inhibitors. Among the segregants from diploid strains of *Aspergillus nidulans*, haploids can have one or other allele at each and every locus originally heterozygous. Diploids, on the other hand, show segregation (becoming homozygous) only for one or a few linked loci, remaining heterozygous for the others. When the original diploid is suitably marked in two or more different chromosome pairs, it is observed that homozygosis affects each chromosome pair independently, one at a time. When a chromosome pair is marked at several different loci, only some of the loci become homozygous in each case. In general, all loci distal to the point of exchange become homozygous, those proximal to this point remaining heterozygous. Illustrative data for an *Aspergillus nidulans* diploid reported by Pontecorvo and Kafer (1958) are given in Table 3.2 in respect of selected white or yellow diploids.

Table 3.2 Mitotic recombination in *Aspergillus nidulans* in respect of linkage groups I(*ad pro1 pab1 y bi*) and II(*w*). Data of Pontecorvo and Kafer (1958).

Original diploid $\frac{w}{+}$ © $\frac{+}{ad}$ © $\frac{pro1 \quad + \quad + \quad bi}{+ \quad pab1 \quad y \quad +}$

Intervals: I (between *w* and ©), II, III, IV, V, VI

Colour of selected segregant	Constitution	Number observed	Interval with recombination
White	$\frac{w}{w}$ $\frac{+\ pro\ +\ +\ bi}{ad\ +\ pab\ y\ +}$	56	I
Yellow	$\frac{w}{+}$ $\frac{+\ pro\ +\ y\ +}{ad\ +\ pab\ y\ +}$	32	V
Yellow	$\frac{w}{+}$ $\frac{+\ pro\ pab\ y\ +}{ad\ +\ pab\ y\ +}$	78	IV
Yellow	$\frac{w}{+}$ $\frac{+\ +\ pab\ y\ +}{ad\ +\ pab\ y\ +}$	7	III

The mechanism of recombination at mitosis is assumed to be by crossing over. Proof requires the recovery of the reciprocal products of one event. The usual methods used for obtaining recombinant products depend on selection for homozygosis and therefore ensure that only one of the recombinant chromosomes is recovered, together with a non-cross over homologue. In special situations it is possible to select for a recombinant without restriction on the composition of the homologue that accompanies it. The method, due to Roper and Pritchard (1955), is to construct a diploid having two different allelic mutants that do not complement, but will recombine. Their diploid had the constitution:

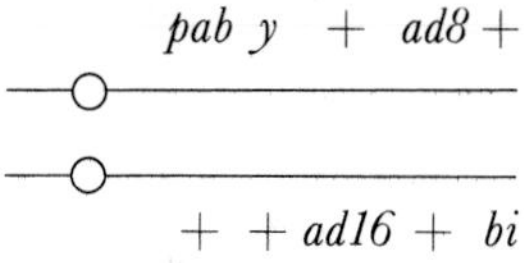

pab y + *ad8* +

—○——————————

—○——————————

+ + *ad16* + *bi*

By recombination between *ad16* and *ad8*, which are allelic, an ad^{+} gene can be generated; this occurs in about one conidium per 10^{7} plated on p-aminobenzoic acid and biotin. The constitutions of the diploids were determined by isolating haploid segregants. In 9 out of 41 cases analysed by Pritchard (1955) the genotype was:

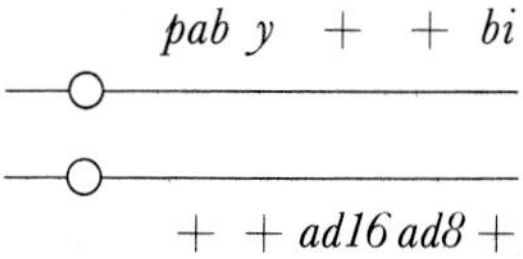

pab y + + *bi*

—○——————————

—○——————————

+ + *ad16 ad8* +

These had two reciprocal recombinant chromosomes, of the constitutions expected from one cross over between *ad16* and *ad8*. The constitution of the *ad16 ad8* double mutant was determined by a laborious programme of outcrossing and selection. The remaining 32 cases had the following genotypes, in respect of the *ad* locus: $27\frac{+\ +}{ad16\ +}$, $4\frac{+\ +}{+\ ad8}$, and $1\frac{+\ +}{+\ +}$. In a number of cases there are multiple exchanges in and about the *ad* locus including instances like $\frac{+\ +\ +\ bi}{y\ ad16 + +}$ that are not readily explicable by plausible classical exchanges. It seems possible that not all mitotic recombinants at the *ad* locus are indeed reciprocal, but in this particular case a substantial fraction are.

Since recombination occurs between alleles at mitosis the question is raised as to whether the process is like that at meiosis involving conversion locally and associated crossing over. Unfortunately, there are some disadvantages. The four products of an event are not available, only half tetrads being recovered in *Aspergillus*. Moreover, the direct products of recombination cannot be isolated; only a population of cells derived from the original event can be examined. A method applicable to cases in which the progeny are recovered from an event occurring in an individual cell, usually induced by radiation, was introduced for yeast by Roman (1963) and will be considered later. It seems not to have been attempted with *Aspergillus nidulans*.

Several other studies have been made of recombination between alleles at several loci in *Aspergillus*, notably by Putrament (1964), Bandiera, Armaleo and Morpurgo (1973) and Jansen (1964) on *pab1*, by Morpurgo and Volterra (1968) on *pfp* (selecting those sensitive to *p*-fluorophenylalanine) and by Putrament (1967) on the two linked loci *ad9* and *pab1*. All these loci are in linkage group I (Fig. 3.2). In every case the parental diploids were marked on

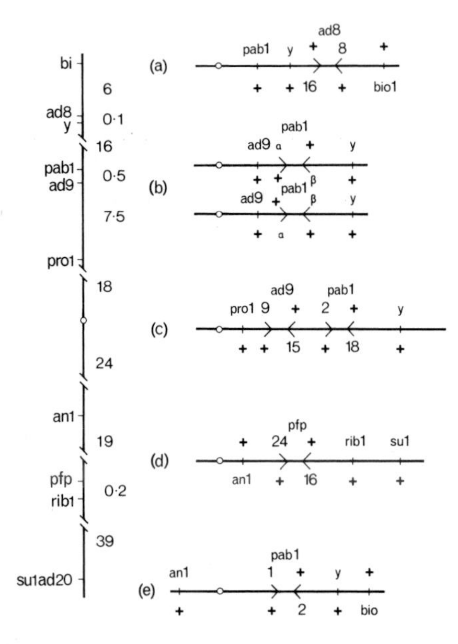

Fig. 3.2 *Aspergillus nidulans*, linkage map of chromosome I showing the loci used in studies of mitotic recombination at several loci. The constitutions of the heterozygotes from which mitotic recombinants were selected are shown as follows: (a) Pritchard, 1955; (b) Putrament, 1964; (c) Putrament, 1967; (d) Morpurgo and Voltera, 1968; (e) Bandiera, Armaleo and Morpurgo, 1973. Symbols are: *su1ad20*, *suppressor of adenine 20*; *rib1*, *riboflavin-1*; *pfp*, *p-fluorophenylalanine* (PFP); *an1*, *aneurin1*; *pro1*, *proline1*; *ad9*, *adenine 9*; *pab1*, *paraaminobenzoate 1*; *y*, *yellow*; *ad8*, *adenine 8*; *bi1*, *biotin 1*.

each side of the locus at which selection was practised; often there were two marked loci on each side. To simplify presentation of the results, consideration will usually be restricted to the closest markers and made more general by designating all diploids as

$\frac{Pm^1 + D}{p + m^2\ d}$. Four classes of prototrophic recombinant are distinguishable: $P+D$, $p+d$, $p+D$ and $P+d$. These may be combined with any one of $+$, m^1+, $+m^2$ or $m^1 m^2$, each in combination with any one of PD, pd, pD and Pd. Thus there are sixteen kinds of genotype for the second chromosome and so 64 possible diploids. The analysis of the diploids that are selected as having a prototrophic phenotype is carried out by selecting haploids from them by growing colonies on a medium containing *p*-fluorophenylalanine (Lhoas, 1961). The diploids are inhibited and vigorous sectors are dependent on the chance occurrence of recessive mutations that are resistant, so haploid segregates are favoured.

The available data are summarized in Table 3.3. Two sets of data, those of Bandiera, Armaleo and Morpurgo (1973) and Morpurgo and Volterra (1968) were not analysed completely. The data of Putrament (1964) provide the most detailed set of $pab1^+$ allelic recombinants, presumably selected at random and analysed completely. Putrament's (1967) data on $pab1^+$ and $ad9^+$ are of unique interest. In this case recombination at two linked loci, *ad9* and *pab1*, 0.5cM apart was followed (1cM = 1% of recombination). Remarkably more than twice as many double ad^+ pab^+ recombinants were obtained when selection was made for pab^+ (4.12%) as when selection was made for ad^+ (1.78%). Coincidence of recombination in the two loci probably occurs much more frequently than by chance, moreover the *ad* and *pab* alleles on the same chromosome convert together. Also, the distal *ad* and *pab* sites convert more frequently than do the proximal *ad* and *pab* sites.

The data show that most recombinants are due to conversion. In Putrament's (1964) data, the doubly mutant *pab* was present with pab^+ in only 20 diploids out of 393 diploids studied; some 196 would be expected if all recombinants were reciprocal. The diploid recombinants mostly have the flanking markers from both parental chromosomes (see below);

	$P+D$	$p+d$	$p+D$	$P+d$
PD	2	99	42	5
pd	142	2	10	30
pD	3	1	0	13
Pd	3	8	32	1
Total	150	110	84	49

homozygosis occurred at the proximal locus in 24 (6.1%) and at the distal locus in 88 (22.4%). The evidence is consistent with most mitotic recombination being due to conversion. Indeed the distribution of flanking markers among prototrophic recombinants is typical of conversion data. The commonest diploids are $P+D/p+m^2d$ and $p+d/Pm^1+D$. Even the two recombinant chromosomes, $p+D$ and $P+d$, are more

Table 3.3 *Aspergillus nidulans*, summary of data bearing on mitotic allelic recombination in diploids at loci in linkage group I. The constitutions of the heterozygotes which generated the recombinants are given in Figure 3.2. Sources of data are: (1) Putrament, 1964; (2) and (3) Putrament, 1967; (4) Bandiera, Armaleo and Morpurgo, 1973, * = not classified for *pab1* alleles; (5) Morpurgo and Volterra, 1968, recombinants are PFP sensitive, † = not classified for *pfp*ʳ alleles, ≠ = semi-resistant to PFP; (6) Pritchard, 1955.

	m^+ chromosome																							
	$P+D$						$p+d$						$p+D$						$P+d$					
Locus	*pab1*	*pab1*	*ad9*	*pab1*	*pfp*	*ad8*	*pab1*	*pab1*	*ad9*	*pab1*	*pfp*	*ad8*	*pab1*	*pab1*	*ad9*	*pab1*	*pfp*	*ad8*	*pab1*	*pab1*	*ad9*	*pab1*	*pfp*	*ad8*
Source of data	1	2	3	4	5	6	1	2	3	4	5	6	1	2	3	4	5	6	1	2	3	4	5	6
Second chromosome																								
$P+D$									1			1												
Pm^1+D	2					1	99	21	25	5*	16+5		42	8	18	5*	19†	16	5					1
$P+m^2D$														1										
Pm^1m^2D																								
$p+d$	2																							
pm^1+d			1			1													4		1			
$p+m^2d$	140	20	11		21†	4	2						10	1		1*			26	8	6	2*		
pm^1m^2																								
$p+D$	2	1	3																3		1			
pm^1+D	1	4	2	1*	2†	1	1		1	1*										13	6		{6† / 2≠}	
$p+m^2D$			1																6		1			
pm^1m^2D																			4					
$P+d$								1	1	2					1									
Pm^1+d	1						5	7	10	3*			10	12	13		{4† / 16≠}	2	1			4*		
$P+m^2d$	2						1	4	2				8	2										
Pm^1m^2d							2					1	14	2				8						

frequently combined with *PD* or *pd* than with the respective complementary reciprocals *Pd* and *pD*. This contrasts with the situation for the *ad8* locus, as reported by Pritchard (1955).

Siddiqui's (1962) analysis of allelic recombination at meiosis in the *pab1* locus shows evidence of a very strong polarity with greater conversion distally (Table 3.4). This indicates a point of initiation distal to the *pab1* locus and suggests that most segments of heteroduplex do not extend from site 18 as far as sites 2, 9 and 5. On the contrary, conversion at mitosis is much less polarized, with some indication of polarity operating from the distal end, though there is evidence of possible initiation from the proximal end. Further, in mitotic recombination it appears that the heteroduplexes may be longer than is characteristic of meiosis.

Table 3.4 Comparison of meiotic and mitotic allelic recombination in *Aspergillus nidulans* at the *pab1* locus with respect to combinations of flanking markers in *pab1*$^+$ recombinants. In the map the spacing of *pab1* sites is in prototrophs per 10^5. Data of (1) Siddiqui, 1962 table 4, *ad9/y*; (2) Putrament, 1964, *ad9/y*; (3) Putrament, 1967 table 6, *pro1/y*; (4) Putrament, 1967 table 5, *pro1/y*.

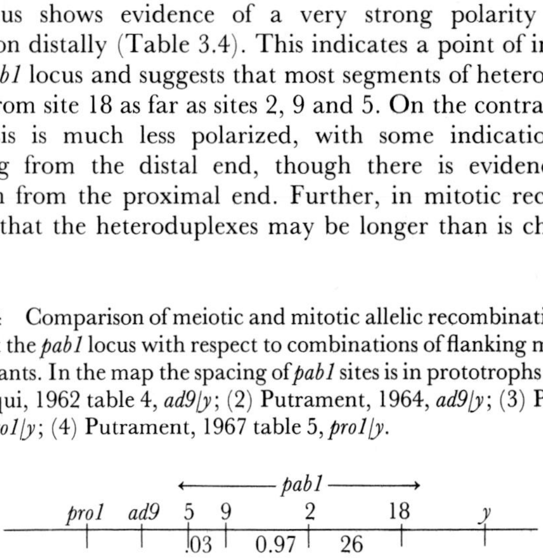

← *pab1* →

pro1 *ad9* 5 9 2 18 *y*

.03 0.97 26

pab1 $m^1 \times m^2$		Meiotic recombination(1) *PD*	*pd*	*pD*	*Pd*	Mitotic recombination *PD*	*pd*	*pD*	*Pd*	Reference
5	2	19	19	59	5	54	29	21	25	(2)
5	18	4	42	68	0	32	25	25	16	(2)
9	18	9	58	110	1	4	7	8	3	(4)
2	18	3	18	33	0	24	33	27	22	(3)

3.2 Saccharomyces cerevisiae

Roman (1956) observed in yeast that diploid cells heterozygous for genes controlling adenine biosynthesis yielded recombinants when dividing mitotically. The method was to use diploids carrying different alleles at the *ad6* locus, comparing *ad6-1* +/+*ad6-2* with *ad6-1*/*ad6-1* and *ad6-2*/*ad6-2*. Whereas *ad6*$^+$ reversions occurred at the rate of 10^{-7} or 10^{-8} in the homozygotes, the frequencies of prototrophs were a hundred or more times greater in the heterozygote. Among more than 100 independent recombinants, the only diploids observed were ++/*ad6-1*+ or ++/+*ad6-2*, usually with a bias towards one of them; no ++/*ad6-1 ad6-2* were observed. Compared with meiosis, only a half tetrad is recovered and it is possible that a bias against the two cross over chromatids going to the same pole at mitosis would explain such a result.

It is possible in yeast to recover all of the mitotic products of a cell in which a recombination has occurred. The original method (Roman, 1963) employed mutants at two different *adenine* loci situated in different linkage groups. Diploid cells homozygous for *ad2* not only need *adenine* for growth, they also produce a red pigment. Mutants at the *ad6* locus are blocked at an earlier step in adenine biosynthesis and the double mutant *ad2 ad6* does not produce red pigment. Thus the diploid *ad2 ad6-1/ad2 ad6-2* requires adenine for growth and forms a white colony. A recombinant at the *ad6* locus resulting in the formation of $ad6^+$ cells would be seen as a red sector in a white colony. Other genetic markers are introduced in the *ad6* chromosome, particularly *trp* and *leu*, which are closely linked and close to the centromere in the opposite arm from *ad6*, and *raf* (fermentation of raffinose or sucrose) and *mal* (fermentation of maltose) that are closely linked and distal to *ad6* in the same arm (Fig. 3.3a).

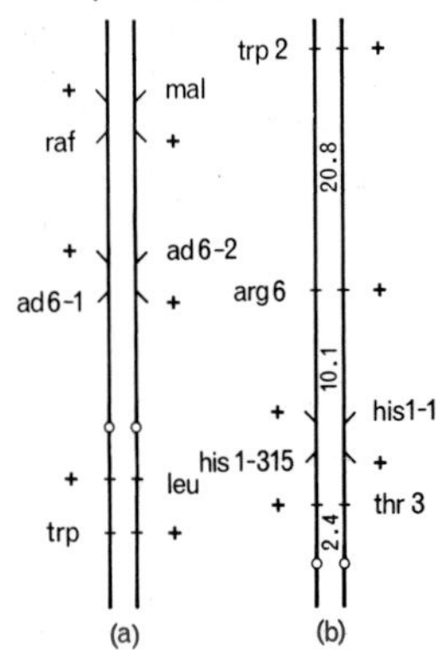

Fig. 3.3 *Saccharomyces cerevisiae* heterozygotes designed for the study of allelic recombination at mitosis in situations permitting recovery of all products of each event. (a) Roman (1963): *trp, tryptophan*; *leu, leucine*; *ad6, adenine 6*; *raf, raffinose*; *mal, maltose*. (b) Wildenberg (1970): *thr3, threonine 3*; *his1, histidine 1*; *arg6, arginine 6*; *trp2, tryptophan 2*.

The heterozygous diploid (yeast workers often describe the situation as heteroallelic) is plated on medium containing adenine and colonies with red sectors are sought. These are rare and the red sectors are often small. The frequency of mitotic allelic recombination can be considerably enhanced by irradiation with ultraviolet light. With relatively low doses, permitting about 90% of the cells to survive, colonies that are half red and half white occur with a frequency around two per thousand, but ranging above and below this value according to the particular pair of *ad6* mutants. A number of cells are isolated from each of the sectors and analysed for their genotypes by inducing sporulation and examining the segregants. The double mutant recombinant is not recovered with the *ad6*$^+$ recombinant. Hence the generation of the recombinants is by non-reciprocal conversion.

While it is evident that allelic recombination at mitosis is mostly by conversion, it appears that the polarity of conversion, found consistently at meiosis, may not occur in spontaneous mitotic events. Wildenberg (1970) studied the immediate products of mitotic cells in which an allelic recombination to yield a prototroph had occurred. The initial heterozygote (marked in linkage group V) is shown in Fig. 3.3b. A culture dividing synchronously was irradiated at the start of a cycle of growth. Cells were plated on medium lacking histidine and prototrophs were detected as those forming a bud. The parent cell, the bud and the next buds of each were isolated and analysed genetically. A total of 129 pedigrees were classified in 68 types; 77 contained only *his*$^+$ cells, 47 had *his* as well as *his*$^+$ and 5 had three or more cells distinct at the *his* locus. The double *his* mutant was present in 9 pedigrees, with the *his*$^+$ recombinant. The flanking marker constitutions of *his*$^+$ chromosomes (more than one in some pedigrees) were 61 *arg6*, 37 *thr3*, 46 *thr3 arg6* and 12++. These data are consistent with the prototrophs being formed mainly by conversion and with a greater probability at the proximal *his1-135* site. Many of the conversions occurred before the chromosome had replicated.

Many agents induce higher frequencies of allelic recombination. These include ultraviolet light (Roman and Jacob, 1958), X-rays (Manney and Mortimer, 1964) and several chemical mutagens (Zimmerman and Schwaier, 1967), such as nitrous acid, 1-methyl-3-nitro-1-nitrosoguanidine and other alkylating agents. The X-ray and chemical mutagen effects provide a basis for fine structure maps of loci. Thus, Manney and Mortimer (1964) showed that the relationship between the frequency of prototrophs among survivors and the X-ray dose was linear and that the slope differed from one pair of alleles to another. The relationship was such that the slope could be used as a mapping metric, with one prototroph per 10^8 survivors per rad equal to one X-ray map unit. Using this metric consistently additive maps can be constructed, e.g. for the *trp* locus concerned with tryptophan synthetase (Manney, 1968). Parker and Sherman (1969) have compared the metric carefully with the fine structure map of the gene for cytochrome c in relation to the polypeptide it specifies and conclude that one X-ray map unit corresponds to 129 nucleotide pairs. Snow and Korsch (1970) have also found a linear relation between prototroph frequency and the square of the exposure time to a given concentration of methylmethane sulphonate (MMS). The X-ray and MMS maps show similar sequences and spacing of mutational sites.

Many agents induce both reciprocal recombination and conversion. If both events are manifestations of a single mechanism, any agent affecting recombination may be expected to do so equally for both. However, Roman (1967) cites evidence that ultraviolet light and mutagenic chemicals (EMS, nitrosoguanidine and hydroxylamine) alter the proportions of the events in different ways. Thus in a diploid AB/ab, with the centromere to the left of the A/a locus, reciprocal recombination is recognized by the recovery of AB/AB and ab/ab cells both of which arise from the same event and are found in sectored colonies grown from treated cells. Non-reciprocal recombination is recognized by the presence of AB/ab with aB/ab cells. Note that in the reciprocal event, the B or b gene is also rendered homozygous, but not so in the non reciprocal event. When ultraviolet light is used as the inducing agent 75% of the events are reciprocal, whereas with EMS only about 5% are reciprocal, most being non-reciprocal.

Although recombination in diploids heterozygous for two alleles is mainly due to conversion, the frequencies provide a reliable metric for mapping. If the event leading to conversion occurs with a constant frequency at a given locus, a considerable proportion of co-conversion should occur, especially when the mutant sites are close together. Fogel and Mortimer (1971) report that when prototrophs are recovered from a diploid of genotype 12++/++34 (where 1, 2, 3 and 4 are sites of mutational difference), the prototrophs are of two classes ++++/12++ and ++++/++34. In both instances, simultaneous conversion of both alleles on the same strand is required.

Mitotic recombination displays many parallels to recombination at meiosis. The question has been raised whether the mitotic events could be due to a small proportion of cells entering into a parameiotic state in which recombination occurs at meiotic frequencies. This is hardly supported by the observation that the average number of mitotic cross overs in a chromosome arm is much smaller than at meiosis. For example, Nakai and Mortimer (1969) show that even in those few cells in which recombination had occurred in VIIL, nearly all progeny could be accounted for by only a single cross over; the meiotic length of 186 cM requires an average of 3.72 cross overs. Whether the genetic controls of mitotic and meiotic recombination are shared, and to what extent, will be seen in the next chapter. In any case, although conversion and crossing over at mitosis are dependent on common mechanisms each has some independent mechanisms and so may be caused to vary independently.

4

Genetic control of recombination

'. . . the behaviour of the chromosomes of an organism in the resting nucleus, in mitosis and meiosis, is subject to the control of the genotype.'

C. D. Darlington (1932), *American Naturalist*, **66**, 47.

Genetic factors having quantitative effects on crossing over and conversion are known in several organisms. They may be grouped broadly into two classes. There are genes of general effect, responsible for the successive stages of recombination, mutations of which are generally recessive and result in the reduction or elimination of recombination throughout the genome. Some of these are known by their cytological properties and are considered in Chapter 5. Secondly, there are genes of local effect, reducing or enhancing recombination in short and scattered parts of the genome.

4.1 Genes of general effect

Recessive mutants of genes of general effect, which virtually eliminate recombination, would be expected to result in high levels of sterility in sexual reproduction. Thus the formation of ascospores in asci and their proper maturation is dependent upon the regularity of meiosis and segregation and, so, on a balanced haploid set of chromosomes in each of its products. A number of attributes are dependent upon various aspects of the metabolism of deoxyribonucleic acid (DNA). These include replication and mitosis, meiosis and recombination, mitotic recombination, mutation and repair of radiation and chemical damage. To the extent that recombination shares processes with other effects of the metabolism of DNA, mutants that are defective in recombination will also be defective in other properties. To the extent that these are necessary for the continuing life of the organism they will be hard if not impossible to discover.

Some mutants that confer sensitivity to ultraviolet light (van de Putte *et al.*, 1966) and X-rays show reduced recombination. In yeast, mutation at 17 loci leads to sensitivity to ultraviolet light but not to X-rays; 14 of these mutants sporulate normally, but 3 show reduced sporulation; spontaneous mitotic recombination is normal in all 6 tested (Snow, 1968). Mutation at 8 loci leads to sensitivity to both ultraviolet light and to X-rays; sporulation is normal in 3 of these, absent in 2 and untested in 3;

performance in recombination is unknown. Mutation at two loci leads to sensitivity to X-rays, but not to ultraviolet light; the mutants will not sporulate and have enhanced spontaneous mutability. A summary source of information is given by Catcheside (1974).

Neurospora crassa

In *Neurospora crassa*, Schroeder (1970) showed that mutation at two loci, of six at which mutation leads to radiation sensitivity, causes infertility. These are *uvs-3* and *uvs-5*. When either mutant is homozygous in a cross, asci do not develop beyond the stage of ascogenous hyphae. In contrast, Smith (1975) has reported a recessive mutant, *mei-1*, which produces abundant ascospores, 90% of which are aborted. Most asci have 0 black:8 white spores, with occasional 2:6 and 4:4. The absence of 8:0 and 6:2 is consistent with meiosis being defective in the first division. Many of the viable spores are disomic, often for several chromosomes, indicating that the inviability of ascospores is caused by non-disjunction at meiosis presumably of the unassociated chromosomes. The disomic cultures rapidly become haploid and in these there is evidence of an absence of recombination wherever a linkage group was marked at several places. Cytologically there is a complete absence of chromosome pairing in crosses homozygous for *mei-1*. Evidently, a defect in at least one stage of meiosis allows development which is prevented by *uvs-3* and *uvs-5*; possibly the latter are unable to enter upon meiosis at all.

Podospora anserina

In *Podospora anserina*, Simonet and Zickler (1972) have found mutants at three loci (*mei1*, *mei2* and *mei3*) in which the prophase I of meiosis is visibly abnormal. At the *mei2* locus one allele is 'leaky' and, in self crosses or in crosses with either of two other alleles, some perithecia containing a few asci are produced. Analyses (Simonet, 1973) of these asci from crosses genetically marked for three linkage groups disclose some changes in recombination frequencies, including second division segregation. Increases are found near centromeres and decreases elsewhere. The observations are difficult to interpret since the asci are selected survivors and to survive must by chance have had a sufficiently normal meiosis to permit regular disjunction. *Podospora anserina* is a secondarily homothallic species. The mating type alleles segregate at the second division in nearly every ascus. After the third division two non-sister nuclei are included in each of the four ascospores. Some sorts of non-disjunction could be compensated in consequence.

Aspergillus nidulans

Fortuin (1971) and Jansen (1970) have reported on a number of mutants of *Aspergillus nidulans* that are sensitive to ultraviolet light. These occur at five loci *uvsA*, *B*, *C*, *D* and *E*. Various experimental tests suggest that *uvsB* and *uvsD* mutants are deficient in excision repair (removal of thymine dimers induced by ultraviolet light) while *uvsC* and *uvsE* may be defective in recombination repair. The B and D mutants show enhanced spontaneous mitotic recombination between alleles, whereas *uvsC* and *uvsE* diminish mitotic recombination; D × D and E × E crosses are virtually sterile.

Ustilago maydis

Some mutants of *Ustilago maydis* that are sensitive to radiations are defective in various aspects of recombination (Holliday, 1967). The most extreme is *uvs2*, in which meiosis is completely blocked, breaking down in the early stages of meiosis; crossing over and conversion are normal in *uvs1* and *uvs3*. Mitotic recombination is reduced or inhibited in *uvs1* and *uvs2*, especially if induced by irradiation. However, the quantitative relations are hard to interpret because of confounding with the lethal effect of irradiation. Mitotic segregation in diploids, presumably due to the loss of distal segments of chromosomes is high in *uvs1*, perhaps due to degradation of DNA chains following excision of thymine dimers.

Mutants defective in recombination have also been sought among those that are deficient in nucleases. In *Ustilago maydis* (Holliday and Halliwell, 1968; Badman, 1972) mutants defective in endonuclease were sought as colonies unable, or less able, to digest DNA present in the medium on which they were grown. By this means *nuc1* mutants concerned with extracellular nuclease were found. Later, by a modified procedure, *nuc2* mutants concerned with an intracellular nuclease were isolated. The two loci are linked with 14–26% recombination; neither has any strong effect on radiation sensitivity. Meiotic gene conversion and mitotic allelic recombination at the *nar* locus, the only one tested, was abolished in the double *nuc1* and *nuc2* mutant. Meiotic crossing over was normal in the one interval tested. Similar mutants at two loci in *Neurospora crassa* (Hasunuma and Ishiwawa, 1972), selected as unable to use DNA or RNA as a sole source of phosphate, do not have any effect on recombination either between or within loci.

Unrau and Holliday (1970; 1972) have sought mutants of *Ustilago maydis* that are defective in DNA metabolism among those unable to grow at an elevated (restrictive) temperature. Five were found, among more than 400 screened, to have a reduced capacity to incorporate ^{14}C-adenine into nucleic acid. None were blocked completely. At the restrictive temperature one (*tsd1*), which is not allelic to the other four, produced long uninucleate filaments which die exponentially after 4 hours temperature treatment. The phenotype resembles that of bacteria starved of thymine, but the mutant could nevertheless be blocked at a step in nuclear division other than DNA synthesis. The latter probability is the more likely because, on recovery from inhibition by the high temperature, allelic and non-allelic recombination is increased and aneuploids are produced. Although these effects seem to be mutually contradictory, it is clear that gene conversion either occurs during the block or a precursory event is established.

Saccharomyces cerevisiae

Several promising steps have been taken with yeast to seek mutants concerned specifically with recombination, both at meiosis and at mitosis. Roth and Fogel (1971) have developed a system whereby mutants defective in recombination at meiosis may be selected. It has the advantage of detecting any mutants even if recessive, as is expected for most, on all chromosomes except chromosome III. A strain disomic for chromosome III and otherwise haploid is heterozygous simultaneously at the *mating type* locus and at the *leucine-2* locus, for *leu2-1* and *leu2-27*. The *mating type* genes are essential for the induction of meiosis, the *leu2* genes to assay recombination at the allelic level. After treatment with ethyl methane sulphonate, 940 clones yielded 91 presumptive mutants with reduced recombination at the *leu2* locus, after meiosis had been induced in cells by transferring them to acetate medium. The apparent mutants grouped into four arbitrary classes according to the number of *leu*$^+$ prototrophs observed in a standard culture compared with parental controls. They are: (1) 5 with no prototrophs; (2) 10 with 10% or less; (3) 55 with 10–50%; and (4) 21 with 50% or more. A further 34 mutants showed simultaneous reduction in both meiotic and mitotic recombination. No mutants with increased levels of recombination were recorded.

The five mutants, in which recombination is completely absent, have been shown (Fogel and Roth, 1974) by complementation tests to be all at different loci and all are recessive to normal. They have been examined further with respect to allelic recombination in diploids (rather than disomics), the initiation and completion of DNA replication precursory to meiosis and the formation of ascospores. Two of the mutants failed to undergo the special pre-meiotic synthesis of DNA following induction by sporulation medium. The other three (*con1*, *con2* and *con3*) completed pre-meiotic synthesis of DNA, but formed no recombinants at the *leu3* locus after induction. All showed some sporulation, 0.4% in *con1*, 5.6% in *con2* and 24% in *con3*, but the ascospores are inviable, presumably due to unbalanced sets of chromosomes. There is evidence of the failure of a meiotic process necessary to normal segregation, a different function in each, but the evidence available does not yet identify the missing function.

Mitotic allelic recombination induced by radiation is subject to genetic control, as Rodarte-Ramón and Mortimer (1972) have shown for yeast. Mutants were isolated in cells disomic for chromosome VIII but otherwise haploid, and heterozygous at the *arg4* locus, namely *arg4-2*/*arg4-17*. Colonies were grown up on solid complete medium from cells that had been subjected to mutagen. They were then replicated to solid synthetic medium lacking arginine and then irradiated with X-rays (2.5×10^3 rad). The dose of irradiation is sufficient to produce about 20 prototrophs per colony in normal strains. Colonies without reversions were saved, from the original plates, as potentially deficient in recombination and tested further. Ten mutants were found and seven were studied in detail. They showed, by complementation and other tests of allelism, a minimum of four loci: *rec1*, *rec2*, *rec3* and *rec4*. Two mutants are sensitive (killed more readily) to X-rays (one being *rec2*), one is sensitive to X-rays and ultraviolet light, while the other four are insensitive to UV and to X-rays. Besides the lack of induction of mitotic recombination by X-rays all, except *rec2*, show a similar lack of induction by UV. There are differences between these mutants and the wild type with respect to meiotic recombination (Rodarte-Ramón, 1972). Meiosis is abortive in homozygotes *rec3* and nearly so in *rec2*. In *rec1*, spontaneous and induced allelic recombination at the *arg4* locus at meiosis is about equal to the control, but in *rec4* it is depressed. Non-allelic recombination appears to be normal in *rec1* and *rec4* strains. Evidently some, but not all, steps in mitotic and meiotic recombination are in common.

Lemontt (1971a) has selected mutants at three loci (*rev1*, *rev2* and *rev3*) that are defective in induced mutation. One locus (*rev2*) is the same as a previously known locus (*rad5*) at which mutation leads to sensitivity to ultraviolet light. Recombination at meiosis in *rev* homozygotes occurs (Lemontt, 1971b) at control frequencies between *leu1* and *trp5* and between *arg4-6* and *arg4-17*. The frequency of mitotic recombination induced by UV or by X-rays, measured for the centromere to *ade2* segment and for *arg4-6/arg4-17* increased more sharply with radiation dose in *rev rev* diploids than in ++ ones. Mitotic recombination, although induced by UV damage, is not correlated with UV mutagenesis. Spontaneous mutability is also under genetic control. von Borstel, Cain and Steinberg (1971) showed by a specially designed fluctuation test that spontaneous mutation rates were enhanced in three different radiation sensitive mutant strains. However, the enhancement was specific to particular kinds of mutation. Three classes of mutant have been distinguished in this work: (1) super suppressor genes of class I (Hawthorne and Mortimer, 1968) that generate mutants which modify tyrosine tRNA so that it recognizes the ochre codon and so suppress ochre nonsense mutants; (2) the nonsense ochre mutants; and (3) frame shift mutants. Two of the radiation sensitive mutants (*rad18* and *rad53*) confer enhanced mutation rates on the super suppressor genes of class I while *rad2* depresses their mutation rate. The radiation sensitive mutant *rad51* enhances the mutation rate of the ochre nonsense locus as well as of the suppressors, but much more in the latter. Moreover, *rad18* and *rad51* also increase mutation rates of a frame shift mutant. It is believed that the mutators act by affecting the addition or deletion of bases in DNA. The connection with recombination is that two of the radiation sensitive mutants are defective in sporulation. The mutatory activity may be due to the mutant genes being less active in detecting and correcting errors in DNA of the types likely to be present in heteroduplexes of hybrid origin.

By an enrichment method, von Borstel *et al.* (1973) have succeeded in isolating mutants that enhance mutation rates. Only a few of these are particularly sensitive to radiations, so most are unlikely to be like those selected by the sensitivity method. Two loci, *mut1* and *mut2*, respectively with two and five alleles, have been identified. Alleles show different types of mutator activity, no doubt depending upon the particular change made in the normal gene. Table 4.1 is a synopsis of mutants in yeast that affect recombination and related functions of DNA metabolism. As indicated already, they have been selected by a wide variety of methods. However, the range of properties of each has been imperfectly determined and no evidence is at hand to show whether or not those selected by different methods are allelic.

Table 4.1 *Saccharomyces cerevisiae*. Synopsis of mutants that affect recombination and related functions of DNA metabolism. Abbreviations: UV, ultraviolet light; XR, X-rays; SPOR, sporulation; SMUT, spontaneous mutation; IMUT, induced mutation; SREC, mitotic recombination; MREC, meiotic recombination; S, more sensitive; N, normal; 0, none; +, increased; −, decreased. References: (1) Catcheside, 1974; (2) Fogel and Roth, 1974; (3) Rodarte-Ramón and Mortimer, 1972; (4) Lemontt, 1971a, b; (5) von Borstel *et al.*, 1973.

Locus	UV	XR	SPOR	SMUT	IMUT	SREC	MREC	Reference
rad5 (=*rev2*)	S	N	N	−	0	N		1
rad18	S	S		+				1
rad51	N	S	0	+				1
rad53	N	S	0	+				
con1			0				0	2
con2			−				0	2
con3			−				0	2
rec1			N?			0	+	3
rec2		S	0			0		3
rec3			0			0		3
rec4			N?			0	−	3
rev1					0		N	4
rev3					0		N	4
mut1	N?	N?		+				5
mut2	N?	N?		+				5

Another method of obtaining mutants of genes affecting recombination is to examine mutants that are defective in sporulation. Bresch, Müller and Egel (1968) applied the method to fission yeast. Although normally heterothallic, this species has a third allele, h^{90}, which promotes homothallism. Colonies of diploid h^{90} cells spontaneously enter meiosis and sporulate after some vegetative growth. Sporulation is detected by exposure to iodine vapour, colonies with many asci becoming blue, whereas those not sporulating remain yellow. Confining the selection among the latter to those that show conjugation, some 300 mutants, mostly recessive, were isolated. They belong to 26 complementation groups. These comprised one locus (*fus*) showing failure of fusion of haploid nuclei; three blocked in meiosis I, together with three others all closely linked to the *h* locus and concerned with commitment to meiosis; one blocked at meiosis II; and 18 blocked at later stages, preventing ascospores being formed around the four haploid nuclei.

Esposito and Esposito (1974) have summarized work on conditional mutants defective in sporulation at the restrictive temperature of 34°C. Both recessives and dominants have been isolated, eleven loci having been identified amongst recessive *spo* mutants. Complementation studies suggest that genes for 50 functions may be recovered. It is possible that some *spo* genes are concerned with functions involved in mitosis, a question that could be settled by testing them for allelism to the conditional mutants defective in the cell division cycle described by Hartwell, Culotti and Reid (1970). Four recessives (*spo7*, *8*, *9* and *11*) and one dominant (*spo98*) fail to undergo the pre-meiotic DNA synthesis after induction by acetate medium. Following induction, increased allelic recombination and crossing over can be detected before the cells are finally committed to meiosis. This property is retained in three mutants (*spo1*, *2* and *3*), all of which fail at some stage of meiosis. Thus commitment to recombination, at meiotic frequencies in some intervals at least, does not commit yeast cells to the reductional division of meiosis nor to the doubling of the spindle bodies required for the first division of meiosis.

Drosophila melanogaster

In *Drosophila melanogaster*, crossing over is confined to females. A considerable number of mutants, beginning with *c(3)G* (Gowen, 1933), affecting their meiosis and recombination have been found, chiefly by testing for increased non-disjunction. So far 21 loci have been identified among 37 mutants (27 by deliberate search, 10 by chance) so far enumerated (Sandler and Lindsley, 1974). Mutants affecting the fidelity of meiosis in the male are relatively much rarer, for in tests of magnitude comparable to those that gave 27 female mutants only four male mutants at three loci have been found; three further loci have been found by chance.

The array of effects seen in female meiotic mutants suggests that the genetic controls of meiosis and mitosis differ mainly, at least, in those genes that are concerned with the events specific to meiosis I, namely pairing, crossing over and segregation of homologues. All known meiotic mutants affect only these properties. The arrays observed include: (1) abolition of crossing over and increased non-disjunction at anaphase I (2 loci, *c(3)G* and *mei-W68*); (2) decrease of crossing over and increased non-disjunction at anaphase I (9 loci); (3) normal or lowered crossing over and increased non-disjunction at anaphase I (5 loci); (4) normal crossing over and increased non-disjunction at anaphase I (1 locus, *mei-S332*); (5) increased crossing over (2 loci). The meiotic mutant, *c3G*, that eliminates crossing over does not affect mitotic recombination (Le Clerc, 1946). However, it does confer increased sensitivity to X-rays (Watson, 1969, 1971). Meiosis in males differs only, it seems, in meiosis I. All mutants, except *mei-S332*, affect only the female. The exception results in precocious separation of sister centromeres in meiosis and consequent equational non-disjunction in both sexes. It is therefore an abnormality in a process unique to meiotic anaphase I that occurs in both sexes; only in that division do sister centromeres normally not separate.

In *Drosophila annassae*, which is closely related to *D. melanogaster*, meiotic crossing over occurs in males (Hinton, 1970). It is dependent upon genes at three or more loci, which show variation in natural populations. Males homozygous for recessive genes at two of these loci and having at least one dose of the dominant at the third locus show levels of crossing over about 10% of that in females. All other genotypes show no crossing over in males and all genotypes show similar values in females.

4.2 Genes of local effect

Compared with procaryotes, eucaryotes show much less recombination per unit of DNA in the haploid set (Table 4.2). This reduction must mean that eucaryotes have acquired controls which considerably reduce, by several orders of magnitude, the probability of recombination. The purposes of control are probably several, but the basic original one was probably the economy of metabolism effected by reducing the action of nucleases. There appear to be three kinds of control, each acting locally rather than generally.

Table 4.2 Relation of recombination frequency and pairing to DNA content and length in various organisms.

Species	Map length	Nucleotide pairs in DNA: per haploid nucleus	Nucleotide pairs in DNA: per map unit	Total DNA	Chromosome length: Zygotene-pachytene	Chromosome length: % paired
Lambda phage	70	4.65×10^4	6.6×10^2	12.3 μm		
T4 phage	800	2×10^5	2.5×10^2	53 μm		
Escherichia coli	2000	10^7	5×10^3	3.7 mm		
Yeast	3700	2.2×10^7	6×10^3	8 mm		
Neurospora	500	4.3×10^7	8×10^4	16 mm	50 μm	0.3
Drosophila	287	2×10^8	7×10^5	61 mm	110 μm	0.2
Maize	1350	8×10^9	6×10^6	250 cm	550 μm	0.02
Man	3300	3×10^9	1×10^6	100 cm		

In eucaryotes where the chromosomes can be observed at the prophase of meiosis, they are considerably condensed before the pairing of homologues occurs. Moreover, the partially condensed homologues do not approach very closely in the synaptinemal complex (see Chapter 5). A total gene by gene matching at zygotene therefore appears to be excluded and it may be inferred that intimate synapsis is restricted to special pairing regions of the chromosomes and that these are the sites of initiation of recombination. These pairing regions might either remain exposed in the partially condensed chromosomes or be subject to exposure for pairing by a specific mechanism. In either case, the pairing regions must have special properties residing in the sequence of nucleotides in their DNA. However, these sequences need not all be exactly alike. Differences between pairing regions in different parts of the chromosomes may influence probabilities of initiation of pairing and so the local frequencies of recombination. Further, differences between allelic pairing genes might be expected to interfere with pairing and so reduce recombination below that found in either homozygote, as found in *Ascobolus immersus* by Girard and Rossignol (1974).

Following pairing it is usually assumed that the next step is the nicking by a specific endonuclease of one chain of one or both of the pairs of DNA molecules which have synapsed. The target of the endonuclease would be in or more probably adjacent to the pairing regions. While these promoter genes which are *recognition* genes (*cog*) for the site of action of a recombinase would in general be similar to one another, some variants may be expected. These would lead to differences in probability of initiation of recombination at different places in the chromosomes. Differences between allelic genes would be expected to lead to situations in which one of the homologues could be preferentially nicked. In such situations, a higher frequency of recombination would be found in the heterozygote as well as the variant homozygote. Examples exist in *Neurospora crassa* (Angel *et al.*, 1970) and in *Schizosaccharomyces pombe* (Gutz, 1971b).

A third type of control involves genes which act dominantly in one or two doses to repress or greatly depress recombination. They act specifically at selected places scattered through the chromosomes. At these places there must be specific recognition sites for the action of each respective repressor gene. Three repressor loci are known in *Neurospora crassa* (Catcheside, 1974) and genes with similar properties are reported for *Schizophyllum commune* (Simchen and Stamberg, 1969) and yeast (Simchen *et al.*, 1971).

Neurospora crassa

(a) Repressor (rec) genes

These were first discovered (Jessop and Catcheside, 1965) as genes which controlled the frequency of recombination, by conversion, between alleles at the *his-1* locus in linkage group V. The dominant gene at the *rec-1* locus, in the same linkage group as *his-1* and about 30 centimorgans distally, reduced the frequency of allelic recombination by a factor of about fifteen. No obvious effects on non-allelic recombination in the neighbourhood could be detected; other cases suggest that this is unusual. However, effects on the neighbourhood of *his-1* were evident since the distribution of flanking markers among prototrophic recombinants was strongly affected by the substitution of *rec-1*$^+$ for *rec-1*. The quantitative effects may be stated simply in terms of the standard cross $Pm^1 + D \times p + m^2 d$, m^1 and m^2 being the sites of difference from normal of two alleles, the m^1 site being proximal. The P and p genes are proximal to the m locus, while D and d are distal. Besides a fifteen fold reduction in the frequency of prototrophs, the *rec-1*$^+$ gene changes the ratios of $PD{:}pd$ from 1.5 : 1 in *rec-1* × *rec-1* to 0.55 : 1 in *rec-1*$^+$ × *rec-1* and *rec-1*$^+$ × *rec-1*$^+$ (Thomas and Catcheside, 1969). The effects can be accounted for if most recombination in the *his-1* locus in the presence of *rec-1* originates proximally to *his-1* and if the action of *rec-1*$^+$ is to inhibit recombination originating from this position. It was presumed that *rec-1*$^+$ acts as a repressor of recombination and that it produces a product which has an affinity for *his-1* or a locus proximal to it and so interferes with the initiation of recombination. The residual recombination was presumed to arise from a source distal to *his-1*.

Subsequently, it was found that *rec-1*$^+$ was highly specific and had no effect on allelic recombination at several other loci. However, it does control recombination at the *nit-2* locus. Moreover, repressor genes at other loci were found to be effective in controlling allelic recombination at some of the loci insensitive to *rec-1*$^+$, particularly *his-2*, *his-3* and *am-1*. The next of these repressor genes to be found was *rec-3*$^+$, which has the effect of reducing recombination at the *am-1* locus by factors of ten to twenty-five (Catcheside, 1966). It was observed that *rec-1*$^+$ was specific to *his-1*, while *rec-3*$^+$ was specific to *am-1*; neither had any effect on a number of other loci. Like *rec-1*$^+$, *rec-3*$^+$ appeared to have no effect on non-allelic recombination in the neighbourhood of *am-1*. The specificity of action implied that there were recognition loci (*con*) in or adjacent to the target loci and that these were quite distinct from one another in character. There would be one type specific for the product of *rec-1*$^+$ and a different type specific for the product of *rec-3*$^+$ (Fig. 4.1).

Smith (1966) found a third locus, *rec-2*, with similar properties except that it controlled non-allelic recombination in linkage group IV. It was found through the action of *rec-2*$^+$ reducing recombination between *pyr-3* and *leu-2*, used as flankers in the study of the fine structure of *his-5*, from 23% to 10%. Subsequently, it was found that the whole effect was concentrated in the *pyr-3 his-5* segment and no other target regions were found in a fairly extensive survey. No effect on allelic recombination occurs in the *his-5* locus; whether any occurs in the *pyr-3* locus cannot be determined due to sterility of crosses between *pyr-3* mutants.

The loci *rec-1*, *rec-2* and *rec-3* are all distinct from one another, their locations being shown in Fig. 4.1. Two alleles are known at each of the *rec-1*

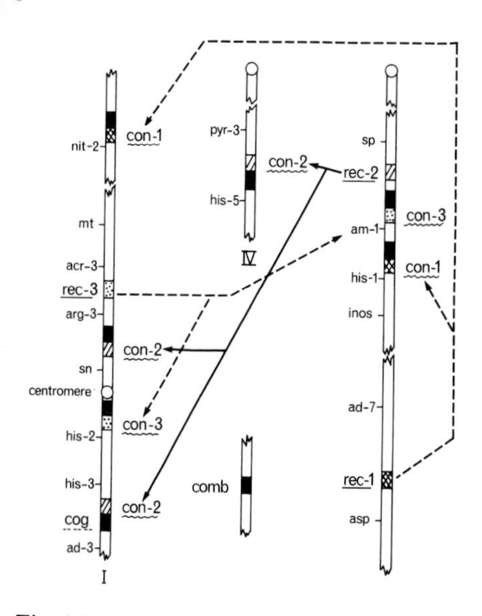

Fig. 4.1 *Neurospora crassa* linkage maps of groups I, IV and V to show the locations of known repressors (or regulators, symbols underlined) of local recombination (*rec-1*, *rec-2* and *rec-3*) and a known promoter (*cog*, dotted underline) as well as the probable locations of inferred *control* genes (analogous to operators, symbols with wavy underline) of three species (*con-1*, *con-2* and *con-3*) and other promoters. Symbols for target and marker loci are: *nit-2*, *nitrate-2*; *mt*, mating type; *acr-3*, *acridine-3*; *arg-3*, *arginine-3*; *sn*, *snowflake*; *his-2*, *histidine-2*; *his-3*, *histidine-3*; *ad-3*, *adenine-3*; *pyr-3*, *pyrimidine-3*; *his-5*, *histidine-5*; *sp*, *spray*; *am-1*, *amination-1*; *his-1*, *histidine-1*; *inos*, *inositol*; *ad-7*, *adenine-7*; *asp*, *asparagine*; *comb*, recombinase. Symbols for *rec* and *con* genes are differentiated according to species.

(Jessop and Catcheside, 1965) and *rec-2* (Smith, 1966; Catcheside and Corcoran, 1973) loci, while three alleles are known at the *rec-3* locus (Catcheside, 1975). In each case, one allele acts dominantly to depress recombination in particular targets, one dose of the dominant allele being as effective as two. The genes at each *rec* locus are specific to a few known target regions (Table 4.3), though there are probably many more. The targets are usually not close to the *rec* locus and are commonly on different chromosomes. The distribution of *rec* genes in laboratory stocks of wild or derivative status (Catcheside, 1975) suggests that the species is polymorphic for these and perhaps other *rec* genes.

It seems probable that each *rec* gene controls allelic recombination at a number of loci and also crossing over in the neighbourhood, probably to one side, of each target locus. This is definitely the case with the following: (1) *rec-3*$^{+}$ reduces allelic recombination in the *his-2* locus and crossing over between *sn* and *his-2* (Catcheside and Corcoran 1973); (2) *rec-2*$^{+}$ reduces

Table 4.3 *Neurospora crassa*, targets of products of regulatory *rec* genes. *Bracketed are number of loci tested in which there is no effect of these *rec* genes on conversion; + = positive action; 0 = no action.

Locus	*rec-1*	*rec-2*	*rec-3*
his-1	+	0	0
nit-2	+	0	0
his-3	0	+	0
am-1	0	0	+
his-2	0	0	+
Others	0(14)*	0(11)*	0(7)*
Region			
pyr-3 his-5	0	+	0
his-3 ad-3	0	+	0
arg-3 sn	0	+	0
sn his-2	0	0	+

allelic recombination in the *his-3* locus and crossing over between *his-3* and *ad-3* (Angel *et al.*, 1970); (3) *rec-1*$^+$ reduces allelic recombination in the *nit-2* locus (D. E. A. Catcheside, 1970; 1974) and crossing over around it. No effects of *rec-1*$^+$ and *rec-3*$^+$ respectively in reducing crossing over near the *his-1* and *am-1* loci have been demonstrated. Possibly the effects are lost within the limits of experimental error. The possible effects of *rec-2*$^+$ in reducing allelic recombination in any of the boundary loci of the *pyr-3 his-5* and *arg-3 sn* segments cannot be examined except in the case of *his-5*, where there are no effects. Allelic crosses between mutants at the *pyr-3* and the *arg-3* loci, respectively, are virtually sterile and at the *sn* locus there is only one mutant gene.

Besides reducing the frequency of allelic recombination at a locus (Table 4.4), a dominant *rec*$^+$ gene also changes, or reduces, the polarity of conversion in a locus. This is manifested in changes in the relative frequencies of the four combinations of flanking markers amongst prototrophic recombinants formed at a locus (Table 4.5). The data given are illustrative. Generally different pairs of alleles at a locus behave similarly, but there are sometimes differences not yet explicable. Commonly there are changes in polarity such as that previously mentioned for the *his-1* locus. Often the changes are small as at the *am-1* and *his-3* (in *cog* × *cog* crosses) loci. There may be quite marked changes due to *rec* genes which affect an adjacent segment, but not the locus itself directly, as with *his-5*.

Any given target region appears to be affected only by the genes at one of the *rec* loci and not by those at the other *rec* loci. This means, as noted previously, that the target regions must each include a recognition gene,

Table 4.4 *Neurospora crassa*, quantitative effects of repressive *rec* genes.
(a) Allelic recombination, measured as prototrophs per 10^5 spores. Loci and alleles: *his-1, histidine-1* (K83, K625); *nit-2, nitrate-2* (MN72, MN73); *his-3, histidine-3* (K504, K874); *am-1*, amination-1 (K314, 47305); *his-2, histidine-2* (K584, K612).
(b) Crossing over, recombinants %. Loci: *pyr-3, pyrimidine-3*; *his-5, histidine-5*; *ad-3, adenine-3*; *sn, snowflake*; *arg-3, arginine-3*.
References: (1) Thomas and Catcheside, 1969; (2) D. E. A. Catcheside, 1970; (3) Angel, Austin and Catcheside, 1970 and unpublished; (4) Catcheside, 1975; (5) Catcheside and Corcoran, 1973; (6) Catcheside and Angel, 1974.

Place of Action		*rec* gene	Crossed to *rec*	*rec*$^+$	Ratio	Reference
(a) Allelic recombination						
his-1		*rec-1*	43	3	14	1
nit-2		*rec-1*	47	10	5	2
his-3		*rec-2 cog*	20	4	5	3
		rec-2 cog$^+$	165	5	33	3
am-1		*rec-3*	25	1	25	4
		rec-3^L	8	1	8	4
his-2		*rec-3*	42	7	6	4
		rec-3^L	7	7	1	4
(b) Crossing over						
pyr-3	*his-5*	*rec-2*	23.3	1.8	13	5
his-3	*ad-3*	*rec-2 cog*	2.6	2.1	1.2	6
		rec-2 cog$^+$	9	2	4.5	6
sn	*his-2*	*rec-3*	6	0.6	10	5
arg-3	*sn*	*rec-2*			2	5

Table 4.5 *Neurospora crassa*. Per cent of prototrophic recombinants in each of the four classes of flanking marker combination.

Loci			*rec* × *rec*				*rec* × *rec*$^+$			
Target	*rec*		PD	pd	pD	Pd	PD	pd	pD	Pd
his-1	*rec-1*		33	22	30	16	21	38	27	15
his-5	*rec-2*		14	33	40	12	7	36	53	5
his-3	*rec-2*	*cog* × *cog*	29	22	30	19	29	15	34	22
		cog × *cog*$^+$	14	32	42	12	29	11	51	9
		cog$^+$ × *cog*	31	13	43	13	29	13	45	12
		cog$^+$ × *cog*$^+$	18	32	31	19	22	28	33	17
am-1	*rec-3*		53	12	19	16	36	25	20	19

consisting of a sequence of DNA, which is able to interact with the products of genes at one *rec* locus, but not with products of those at other *rec* loci. These recognition genes may be referred to as *control* (*con*) genes. There would be a species of *con* genes, each of which could be called *con-1*, *con-2* and *con-3*, corresponding to each species of *rec* locus and there would be a variety of each species of *con* gene situated in each target region.

So far no individual *con* locus has been located by examination of the genetics of a variant gene. However, the probable position of any *con* locus can be deduced for most of the target regions on one or both of two criteria. If the frequency of crossing over is controlled in a segment, but not outside it, by genes at a *rec* locus there must be a *con* locus in the segment. If the frequency of allelic recombination is controlled in a target locus, the *con* locus would be near to the end of the target locus where the highest frequency of conversion is seen in relaxed (*rec*) crosses and where the polarity is reduced or reversed in restricted (rec^+) crosses. These criteria would place a *con-1* locus proximal to *his-1*, *con-2* loci between *pyr-3* and *his-5* and between *his-3* and *ad-3* and *con-3* loci proximal to *am-1* and between *sn* and *his-2*. There would also be a *con-1* locus near to *nit-2* and a *con-2* locus between *arg-3* and *sn* or even to the left of this segment.

The degree of effect of *rec* genes varies considerably as between different targets (Table 4.4). Thus the ratio of effect of *rec-2* versus *rec-2*$^+$ ranges from two in the *arg-3 sn* segment to about thirteen in the *pyr-3 his-5* segment. Even larger differentials are found with allelic recombination. In the case of *pyr-3 his-5* it appears that *rec-2*$^+$ almost completely prevents an event which occurs in the *pyr-3 his-5* segment in nearly all meiotic cells if *rec-2*$^+$ is absent. In *rec-2* × *rec-2* crosses, crossing over between *pyr-3* and *his-5* is about 23.3% compared with 1.85% in *rec-2* × *rec-2* crosses. The absence of *rec-2*$^+$ adds some 21.5 centimorgans to the length of the *pyr-3 his-5* segment. Each meiotic cell in which a cross over, presumably reciprocal, occurs between *pyr-3* and *his-5* would produce an equal number of recombinant and non-recombinant spores. The absence of *rec-2*$^+$ means the occurrence of a chiasma between *pyr-3* and *his-5* in 47% of meiotic cells, compared with 3.7% in its presence. Previously it has been evident that allelic (conversion) and non-allelic (crossing over) recombination are different manifestations of one common mechanism. Moreover, half of the tetrads in which conversion has occurred show correlated crossing over between flanking markers, while the other half do not. It appears probable that an event leading to conversion, as well as crossing over in half of the affected tetrads, occurs between *pyr-3* and *his-5* in 94% of all meiotic cells in *rec-2* ×*rec-2*. These events are stopped in all but 7% of the meiotic cells by the presence of *rec-2*$^+$. The quantitative effects are less extreme in other targets of *rec-2* genes. The observed differences in response suggest that the varieties of *con-2* in the three segments may not be all alike. However, other factors could bring about the different responses.

Evidence for varietal differences of *con-3*, the recognition genes for the products of genes at the *rec-3* locus, is provided in another way. At least three different alleles occur at the *rec-3* locus in the left arm of linkage group I. They are *rec-3* present in Emerson a, *rec-3*$^+$ present in Emerson A and *rec-3*^{L} present in Lindegren A (Catcheside, 1975). The three *rec-3* alleles differ in their action on allelic recombination at the *am-1* and *his-2* loci (Table 4.4). The relative effects of *rec-3*$^+$, *rec-3*^{L} and *rec-3*, each combined with *rec-3*, on allelic recombination are 1:8:25 at the *am-1* locus and 7:7:42 at the *his-2* locus. In each case the gene giving lower yields is dominant to that giving higher yields of recombinants. The *rec-3*$^+$ and *rec-3*^{L} genes are distinguishable by their action at the *am-1* locus but not at the *his-2* locus. The *con-3* gene near to *his-2* evidently does not differentiate between the products of *rec-3*$^+$ and *rec-3*^{L} respectively and must therefore be different in some minor respect from the *con-3* gene, near to *am-1*, which does make the distinction. Non-allelic recombination between *sn* and *his-2* is apparently similar in *rec-3*^{L} crosses to the values in *rec-3*$^+$ crosses.

The failure of the *con-3* gene near to *his-2* to differentiate between *rec-3*$^+$ and *rec-3*^{L} questions the interpretation of negative tests for the possible action of genes at the *rec-1*, *rec-2* and *rec-3* loci on a number of loci. Conceivably, the recessive alleles at each *rec* locus produce products of lower affinity for *con* genes and the absence of a difference due to the dominant versus recessive *rec* genes reflects a failure of differentiation between their products by some *con* genes. Nevertheless, the order of dominance of the *rec* genes is the same with respect to each target. If the recessive did produce a product and if *con* genes did vary, it is conceivable that a reversal of dominance could occur with a *con* gene having greater affinity for the *rec* product than for the *rec*$^+$ product.

The possible dominance relations of alleles at a *con* locus are difficult to predict. If it is assumed that the interaction between a *rec*$^+$ gene and its corresponding *con* genes occurs through affinity between a product of the *rec*$^+$ gene and each of its corresponding *con* genes, there is a close analogy to regulatory gene and operator gene respectively. The association of *rec*$^+$ gene product and *con* gene might act either to restrict pairing in the neighbourhood of the *con* locus or to frustrate initiation of an act of recombination near to the *con* locus after pairing. In the latter case the complex of *rec* product and *con* on just one chromosome might act to prevent the initiation of recombination (say, the nicking of a DNA chain) either in both homologues or only in the homologue with the complex. Thus in a heterozygote for two allelic *con* genes having different affinities for the same *rec*$^+$ product it is predicted that low frequency would be dominant to high frequency, unless the effects were restricted to the homologue with the complex. In the last case low frequency would be recessive to high frequency and initiation and conversion would be restricted very largely to one chromosome, that without or with a lower frequency of the complex. The only satisfactory evidence would come from a case of allelic *con* genes which responded differently to the same *rec*$^+$ gene, since only in this case is it certain that a regulatory substance is produced. There is one instance of a regulatory locus closely linked to *nit-2*, the dominant having the effect of reducing allelic recombination proportionally in both *rec-1* × *rec-1* and *rec-1* × *rec-1*$^+$ crosses (D.E.A. Catcheside, 1974).

(*b*) *Recognition* (*cog*) *genes*

The only locus with this class of gene known definitely is situated between *his-3* and *ad-3* about a quarter of the genetic distance from *his-3* towards *ad-3*. Two alleles occur, their effect on the frequency of recombination being expressed only in the absence of *rec-2*$^+$. In the presence of *rec-2*$^+$ allelic recombination in the *his-3* locus and crossing over between *his-3* and *ad-3* are at the same frequency in *cog*$^+$ stocks as in *cog* stocks. In *rec-2* × *rec-2* crosses the frequency of allelic recombination in *his-3* is about six to eight times as great in *cog*$^+$ crosses as in *cog* × *cog*. Crossing over between *his-3* and *ad-3* is about four times as great in *cog*$^+$ as in *cog* × *cog* crosses. Apparently *cog*$^+$ is completely dominant, for no difference in frequency is seen between *cog*$^+$ × *cog* and *cog*$^+$ × *cog*$^+$. However there are differences in other respects. As is shown in Table 4.5 in the *cog*$^+$ × *cog* heterozygote the *his-3* site in the *cog*$^+$ chromosome is preferentially converted irrespective of whether it is proximal or distal to the *his-3* site in the homologue with *cog*. In consequence *cog* × *cog*$^+$ and *cog*$^+$ × *cog* show a reversal in the ratio of *PD* to *pd* amongst prototrophic recombinants. In contrast, in *cog*$^+$ × *cog*$^+$ the distal *his-3* site is preferentially converted. The behaviour suggests that *cog* is the locus at which, after synapsis, the formation of a heteroduplex DNA is initiated perhaps following nicking by an endonuclease. There is a strong preference for conversion in the *cog*$^+$ chromosome in the *cog*$^+$ × *cog* heterozygote and perhaps a heteroduplex is formed as a rule on only one chromatid of the tetrad.

The interpretation of the function of *cog*$^+$ is reinforced by the behaviour of heterozygotes in which one *his-3* mutant (TM429) is due to an interchange with a break in the *his-3* locus itself (Catcheside and Angel, 1974). This mutant has *cog*$^+$ in the distal segment of linkage group I which is translocated to linkage group VII. In heterozygotes in which *cog*$^+$ is associated only with TM429 no differential effect of *rec-2* and *rec-2*$^+$ is observed for recombination proximal to the site of TM429. If *cog*$^+$ is associated with both TM429 and its structurally normal homologue, the typical large differential effects of *rec-2* versus *rec-2*$^+$ are observed. The action leading to recombination is confined to the chromosome, or chromosomes, which carry *cog*$^+$. When the TM429 interchange chromosome is the only one that has *cog*$^+$, the heteroduplex formation is apparently arrested at the point of interchange and cannot extend to the proximal part of the *his-3* locus.

Genes with properties similar to *cog*$^+$ are present in the proximal part of linkage group I of *Neurospora sitophila*. Fincham (1951) demonstrated a significant difference between its linkage map and that of *N. crassa*. The former showed much higher frequencies of crossing over near the centromere, the difference apparently not extending to more distal parts of the linkage group (Figure 4.2a). Newcombe and Threlkeld (1972) have further shown that a genetic factor (or factors) in the neighbourhood of the centromere of *N. sitophila* acts as a dominant enhancer of crossing over frequencies on both sides of the centromere. The species cross is itself very sterile, but by appropriate back crosses the proximal region of linkage group I of *N. sitophila* was transferred to *N. crassa* and a similar region of *N. crassa* transferred to *N. sitophila*. Crossing over in the centromere region of linkage group I is as high in *crassa* × *sitophila* crosses as it is in *sitophila* crosses (Fig. 4.2b).

Crosses between two strains, A *rec-s* (principally *N. crassa* except for the centromere region of linkage group I of *N. sitophila*) and a *his-2 ad-3B rec-c* (wholly *N. crassa*), where *rec-s* is the genotype for high recombination between *his-2* and *ad-3B* (about 22.1 cM) and *rec-c* the genotype for low recombination (about 1.2 cM) showed the existence of at least two loci in the segment contributing to high versus low frequency (Hargrave and Threlkeld, 1973). One locus, *rec-s2*, is clearly defined at a position 5.4 cM proximal to *ad-3B* (Fig. 4.2c). The other, *rec-s1*, in a region extending

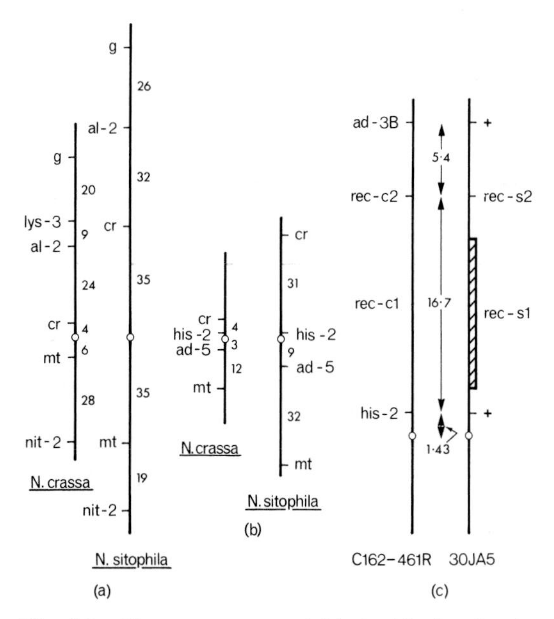

Fig. 4.2 *Neurospora crassa* and *N. sitophila* showing innate differences in recombination frequency in proximal part of linkage group I and location of promoter genes (*rec-s*) transferred from *N. sitophila* to a stock that is mainly *N. crassa*. Data: (a) Fincham, 1951; (b) Newcombe and Threlkeld, 1972; (c) Hargrave and Threlkeld, 1973. Symbols: *g, ginger*; *lys-3, lysine-3*; *al-2, albino-2*; *cr, crisp*; *mt, mating type*; *nit-2, nitrate-2*; *his-2, histidine-2*; *ad-5, adenine-5*; *ad-3B, adenine-3B*; *rec-c* and *rec-s, recombination factors* respectively in *N. crassa* and *N. sitophila*.

distally from near *his-2* towards *rec-s2*, is probably a complex of more than one genetic locus, perhaps several, affecting recombination frequency locally. However, the available data do not permit a further dissection. Like *cog*$^+$ the *rec-s* genes affect recombination in the segments of the chromosome within which they lie. Although *rec-s2* must be close to the site of *cog*$^+$, it is unlikely to be allelic since it is expressed in stocks which most probably carry *rec-2*$^+$.

Schizophyllum commune

Extensive studies of recombination between the systems of genes controlling mating type have disclosed controls similar to those in *Neurospora crassa*. Simchen and Stamberg (1969) proposed a distinction between fine and coarse controls of recombination. Genes of coarse control (equivalent to those which determine the nucleases effective in recombination) are those which have extreme effects on recombination ('all or none') throughout the entire genome, with only rare variants in nature. They occur in procaryotes and eucaryotes and control the steps of synapsis, breakage, repair and separation which must operate sequentially. Genes of fine control are those of small effect on small, highly specific regions of the genome, with variants occurring commonly in natural populations, but only of eucaryotes,† the regional specificity being imposed on at least one step of the coarse control. The implication is that there are two kinds of genes of fine control, one producing the controlling materials and the other constituting the recognition sites situated in the regions in which control is exercised. The theory is expressed that fine control has a function in reducing recombination to a level which allows the maintenance in a population of a high number of mating type factors and so a high potential for outbreeding, the potential for inbreeding being kept low and near to its minimum.

In *Schizophyllum commune* two unlinked systems *A* and *B*, often referred to as two mating type loci, control different aspects of the process of dicaryon formation. Each of *A* and *B* consists of linked genetic loci, with genes of different function, designated α and β and yet other genes can occur between the α and β factors. Complete success in mating between two strains depends upon them being different at the $A\alpha$, $A\beta$, $B\alpha$ and $B\beta$ loci. Identity at some loci permits the formation of a dicaryon to proceed in part. Studies of control of recombination have involved the use of α and β factors in both the *A* and the *B* systems. All stocks are derived from wild sources.

Simchen (1967) was able to select for high (14%) and low (4%) recombination between $A\alpha$ and $A\beta$, the difference being due to genes of major effect occurring at a *rec* locus linked to the *A* system and to genes of minor effect at other loci. Low frequency of recombination was dominant to high. Stamberg (1968) showed that recombination in the *A* and *B* systems was under separate control. Two strains 14(A4 B4) and 699(A41

†The *chi* genes described in lambda phage and *E. coli* by Stahl, *et al.* (1975, *J.M.B.*, **94**, 203–12) are analogous to *cog*$^+$.

B41) were crossed and progeny ('14' and '699') were selected with the mating types of the two parents, eight of each. Each of these progeny was then backcrossed with a compatible parent. Then the recombination between the α and β loci in each of the *A* and *B* systems was measured at 23°C and 32°C in a sample grown from random basidiospores, the sample size ranging from 84 to 196. The identification of the mating type genes depended upon crossing each of these progeny to two testers each being like one parent except for having the $B\beta$ gene of the other parent. The responses allow deduction of whether each progeny is a non-recombinant, a recombinant *A* or a recombinant *B*. The results when analysed for heterogeneity showed: (1) significant variation in the *A* segment recombination at 32° (Fig. 4.3) but not in the *B* segment; (2) significant, but uncorrelated, variation in recombination at 23°C in both *A* and *B* segments; (3) differences between recombination at the two temperatures, the occurrence of differences in the two segments being uncorrelated. Clearly several genes are involved, a minimum of two for each segment, but none affecting both segments. It seems likely that four loci, none linked to the target segments nor to one another, are sufficient to account for the data, taking into account the inherent difficulties and the evident effect of

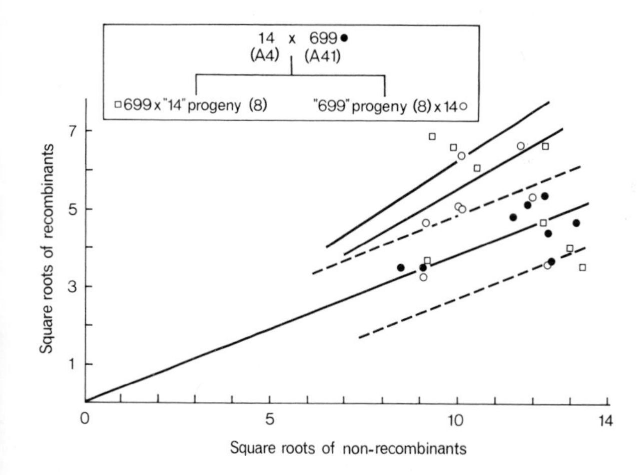

Fig. 4.3 *Schizophyllum commune* recombination between α and β loci in the A mating system measured at 32°C. Eight replicates of the cross between strains 14(A4) and 699(A41), eight A4 progeny of this cross themselves crossed to strain 699 and eight A41 progeny crossed to strain 14 were tested for recombination between the Aα and Aβ loci. Four of the A4 segregates and three of the A41 segregates agree in recombination frequency with the parents (average 11.8%, the lower slope); the remainder show higher recombination frequencies. Four A4 segregates average 27.9% recombination; five A41 segregates average 23.2% recombination. The data (calculated from Stamberg, 1968) are plotted on a square root chart; 95% confidence limits are shown by broken lines either side of the slope representing the parental average.

temperature variation. Thus the eight replicates of the parental recombination values in the *A* segment range from 8.2 to 16.2% (mean 12.9%). Four of the '14' × 699 and three of the '699' fit into this range (Fig. 4.3). Evidently the two parents differ at two loci, 14 being *reca recb*$^+$ and 699 being *reca*$^+$ *recb*. In the presence of *reca*$^+$ and *recb*$^+$ recombination at 32°C is about 13% in the *A* segment; in *reca*$^+$ *recb* it is about 28% and in *reca recb*$^+$ about 23%. Interestingly, the relief from control is not lifted at 23°C as though the actions of *reca*$^+$ and *recb*$^+$ were temperature sensitive. These control genes are of the same character as the repressor *rec* genes in *Neurospora*. A peculiar feature, not further explored, is the appearance at 23°C of a recessive factor which reduces recombination in the *B* segment.

Stamberg and Koltin (1973) present evidence for a genetic difference between strains due to factors located in the recombining region itself. They argue that the segregation amongst the progeny indicates that the difference occurs at a number of sites with additive effects, some positive and others negative. From this and the dominance relationships, with high frequency dominant to low, they suggest that these sites may be recognition sites for the fine control of recombination. They appear to correspond to the *cog* genes of *Neurospora*.

Analysis was made of recombination in the B segment in progeny of the cross 97(Bα6–β7) × 14(Bα2–β6). Some were non-recombinants like the parents. Others were the recombinants (Bα6–β6) and (Bα2–β7). All were measured for recombination in crosses to 699(Bα3–β2) (Fig. 4.4). The difference between the values characteristic of the parental tests (8.0% for strain 97; 5.9% for strain 4) was maintained by the non-recombinant progeny (means 8.1% and 5.3% respectively) and most of the recombinant progeny were also similar. According to Stamberg and Koltin the α6–β6 recombinants were homogeneous, though one had much higher values than the rest (Fig. 4.4d). The α2–β7 recombinants were inhomogeneous with two having very low values and one with a higher value (Fig. 4.4c). *Prima facie* the controlling elements in the B segment do not segregate as one gene.

It is clear that in *Schizophyllum* there are genetic factors of the *rec* type found in *Neurospora*. Three affect the A segment and two the B segment to major degrees. Reduction of recombination in unlinked or closely linked target segments is due to the action of dominant variants. There are also recognition sites within the target segments, but whether these are of the *cog* or *con* type or both is uncertain.

An interesting observation (Stamberg and Koltin, 1971) is that not all pairs of different *B* systems can show recombination. The strains which do not show recombination nevertheless each show recombination with a number of other strains. This is reminiscent of evidence for pairing genes shown in *Ascobolus immersus* (p. 85).

Schizosaccharomyces pombe

Gutz (1971b) has reported a mutant (*ade6*-M26) locus which increases

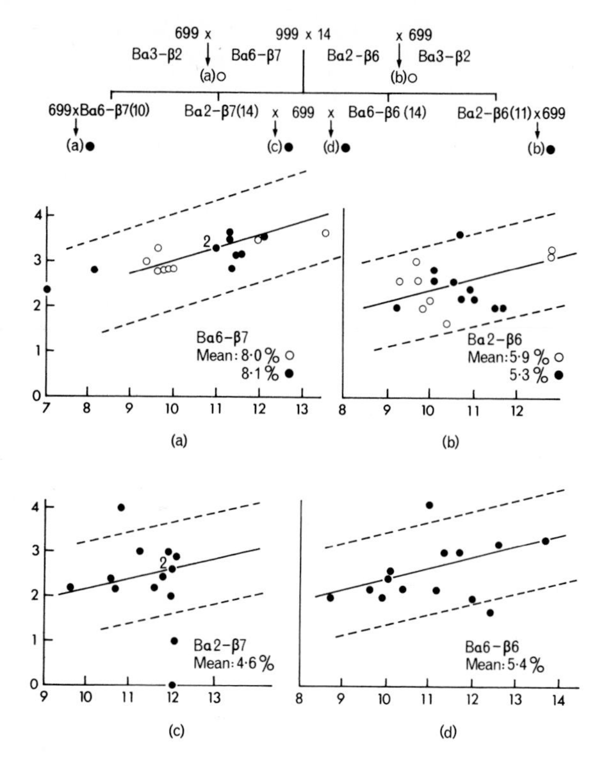

Fig. 4.4 *Schizophyllum commune*. Pedigree and progeny tests for the control of recombination in the B mating system, due to controlling genes within the segment itself. Only progeny recombinant for the α and β genes show substantial differences. The data are calculated from Stamberg and Koltin (1973) and displayed on square root charts. The broken lines are 95% confidence limits.

recombination in the *ade6* locus by large amounts (Table 4.6) of up to twenty times. Moreover, the effects extend to other parts of the locus. Mutation of the *ade6*$^+$ gene results in a requirement of adenine for growth. On yeast extract agar the mutants show colonies pigmented red, whereas the normal wild type is white. Different mutants show different degrees of colour. Thus M26 and M375 are dark red, while M216 and L52 are light red. The colour differences serve to identify at a glance segregations that are not 2:2. In crosses of *ade6*-M26 to the wild type, about 3–5% of asci show conversion and in these M26 is more often converted to wild type than the homologous normal site is converted to mutant, in the ratio of about twelve to one. In heterozygotes between M26 and other *ade6* mutants, the neighbouring sites M216, L52 and M375 also show conversion with nearly 100% co-conversion for M216 and about 60% for

Table 4.6 *Schizosaccharomyces pombe*. Recombination at the *ade6* (*adenine-6*) locus showing the effect of mutant allele M26 in enhancing the frequency. Data of Gutz, 1971b.

M216 M375 M26 L52 M210

Map of mutant sites

Cross	Convert asci %	Prototrophic recombinants per million spores
M216 × +	0.4	—
M375 × +	0.9	—
M26 × +	5.1	—
L52 × +	0.3	—
M216 × M375	—	19
M216 × M26	3.1	68
M375 × M26	—	0.9
M375 × L52	—	314
M26 × L52	3.6	3620

Table 4.7 *Schizosaccharomyces pombe*. Types of asci with converts at the *ade6* (*adenine 6*) locus showing the influences of the M26 mutant stock in biassing the direction of conversion and in promoting co-conversion. Besides needing adenine for growth, M26 and M375 form dark red colonies, M216 and L52 light red colonies, while wild type forms white colonies. These allow identification at a glance of segregations that are not 2:2. The numbers at the head of each column record white : light red : dark red progeny from each tetrad. Data of Gutz, 1971b.

Cross	3:0:1	1:0:3	3:1:0	1:3:0	0:3:1	0:1:3	1:2:1	1:1:2	Others
M26 × +	46	6							
M216 × +			4	3					
L52 × +			2	1					
M375 × +			3	6					
M216 × M26					29	3			1
M26 × L52					19	2	13	1	3

the more distant L52 (Table 4.7). Moreover, the direction of conversion at the M216 and L52 sites is commonly opposite to that at the M26 site. Gutz ascribes the effects to the M26 site itself causing an additional point of preferential breakage within the *ade6* locus. A closely linked gene with the properties of a *cog*$^+$ gene would explain the observations, especially if it had the effect of producing a heteroduplex in one chromatid only, namely the one carrying M26 and the presumed control gene. Gutz cites as evidence against a gene outside the *ade6* locus the observation that converts to M26, M216 and L52 retain the conversion frequency characteristic of the original mutant. For example, in the cross presumed to be *cog* M216 × *cog*$^+$ M26, the converted asci would be 3 *cog* M216 : 1 *cog*$^+$ M26 or

1 *cog* M216:3 *cog*$^+$ M26, the genes at the *cog* locus being co-converted with the *ade6* gene to which each is closely linked.

Ascobolus immersus

Emerson and Yu-Sun (1967) observed that the two wild strains, K5(+) and P5(−), constituting the Pasadena strains of this fungus differ in respect of a number of genes affecting the frequency of conversion at some loci concerned with ascospore colouration and also the relative frequencies of the classes of abnormal segregation constituting the segregation spectrum. Evidence was found of close linkage of genes controlling the frequency of conversion to the locus in which conversion is expressed. A locus close to *w62* is similar to those described later by Girard and Rossignol (see below), in that homozygosity for the gene in K5 or in P5 allows frequent conversion at the *w62* locus, whereas heterozygosity for the K5 and P5 alleles reduces conversion.

A different kind of effect is due to a gene derived from P5 and closely linked to *w10* which brings about a considerable increase in conversion frequency only when coupled with *w10* and not when in repulsion phase (Emerson and Yu-Sun, 1967). Thus the crosses *w10*(K) × +(P) and *w10*(K) × +(K) show low frequencies, with about 2% 6+:2*w*, while *w10*(P) × +(K) shows nearly 4% and *w10*(P) × +(P) shows about 12%. This is unlike any other genetic control so far analysed. The effect is accompanied by changes in the relative frequencies of the four usual types of ascus, but whether this is due to the same gene is not established.

Some effects on the frequency of recombination at the *b2* locus have been mentioned previously, but the most interesting and novel effects are those reported by Girard and Rossignol (1974). They found four distinct factors *cv1*, *cv2*, *cv4* and *cv6* (*cv* for conversion modification) each controlling conversion respectively at the *b1*, *b2*, *b4* and *b6* loci, mutation at each of which leads to colourless ascospores. There appear to be two alleles known at each of the *cv* loci, each present in a different wild strain. The two alleles may be distinguished as *A* and *B*, thus *cv2A* and *cv2B*. In crosses of *b2* × +, the frequency of conversion is high in *cv2A* × *cv2A* and *cv2B* × *cv2B*, but low in the heterozygote *cv2A* × *cv2B*. This relation holds for all three of the *cv1*, *cv2* and *cv4* systems and it seems likely that *cv6* acts similarly. No recombination between *cv1* and *b1* nor between *cv2* and *b2* has been observed, so the respective loci must be close together.

The *b4* and *b6* loci are linked. When both *cv4* and *cv6* are heterozygous the frequency of recombination between *b4* and *b6* is 13.6 cM. Recombination between *cv4* and *cv6*, measured on 50 progeny, is the same at 12 cM. The heterozygous state at each of the *cv4* and *cv6* loci reduces recombination between *b4* and *b6* from 34.8 cM. If only one of the *cv* loci is heterozygous the value is intermediate, 21.3 in one case and 23.7 cM in the other.

These properties are exactly those expected of pairing genes. Intimate molecular pairing would require precise, or almost precise, identity between the segments. Marked molecular differences would inhibit pairing and so not allow any of the succeeding events necessary to recombination to be instituted effectively. Hence heterozygosity of pairing genes would depress conversion and crossing over locally.

5

Meiosis

For all that moveth doth in Change delight.
Spenser, *The Faerie Queene*

From its onset meiosis differs from mitosis in that the chromosomes appear to be single when they first become visible at leptotene. Although their DNA, or nearly all of it, replicated during the S-phase of the preceding interphase, the duality is not reflected visibly as separate chromatids. The centromeres of the chromosomes are usually polarized and the ends of the arms are attached to the nuclear membrane according to studies by electron microscopy. Homologous threads now pair (at zygotene) in a highly specific manner, such that all morphological features (chromomeres, centromeres, nucleolar organizers, knobs) are brought into juxtaposition from end to end. In a diploid, the chromosomes are grouped into a haploid number of pairs. The stages from leptotene to diplotene are illustrated in Fig. 5.1.

During pachytene the chromosomes remain intimately paired and they shorten and thicken and twist around one another to form relational spirals, as though they were anchored together bifacially. However, this can hardly be a direct function of the DNA duplex. After a time, the homologues of each pair separate, but incompletely. Where they separate, each homologue is seen to consist of two closely apposed chromatids. At one or more places along the length of the bivalent (pair) the chromatids of each homologue change partners. Each of these visible points of exchange is a chiasma, where two chromatids lie crossing over one another. The mechanism of their formation cannot be observed directly, but it is possible to show that some chiasmata are the result of a physical exchange of material between two chromatids, as though each is broken at a particular point and the parts rejoined in new ways. It is generally supposed that all chiasmata are the result of actual exchange of material. The relation of chiasmata to recombination is discussed by Henderson (1970). The formation of chiasmata is accompanied by a reduction in the number of relational coils and it was at one time supposed that the coiling provided energy for breaking the chromatids. The chiasmata serve mechanically to retain homologues in associated pairs, although elsewhere the homologues appear to repel one another. Further contraction results in compact

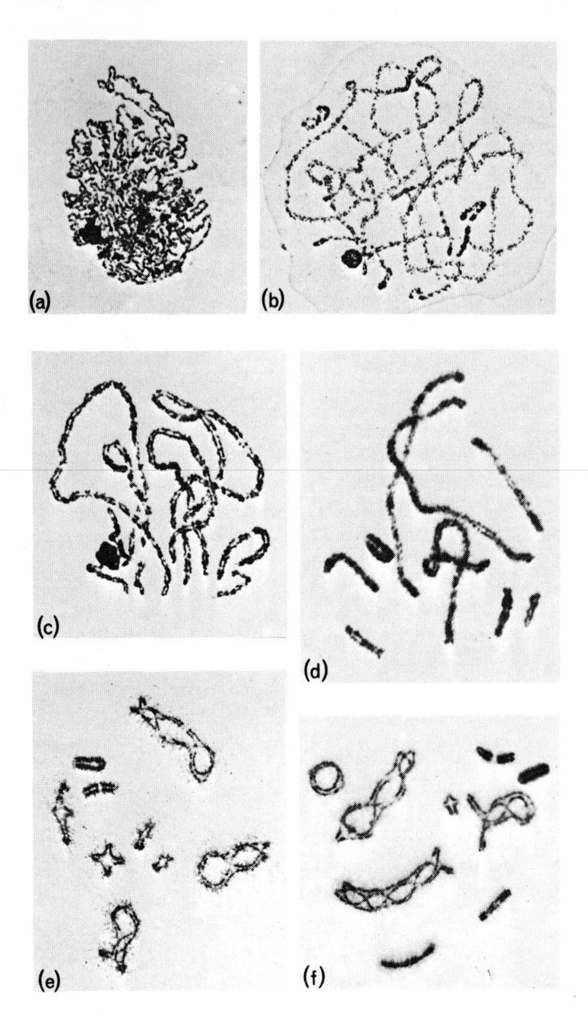

Fig. 5.1 First prophase of meiosis in the male meadow grasshopper, *Chorthippus parallelus* (×720). This species has a diploid chromosome number of 17, the uneven count stemming from the presence of a single sex (X) chromosome. This behaves as a univalent element throughout the entire first division of meiosis and during first prophase appears as a condensed and deeply staining structure. The sixteen autosomes, which range in size, are unpaired at the onset of meiosis when they form a tight cluster of fine threads (leptotene, a). They subsequently become associated as eight bivalents which, initially, are extended and two stranded entities (zygotene, b and c) arranged in a polarised fashion. These condense, lose their polarity (pachytene, d) and open out into four stranded bivalents in which pairs of non-sister chromatids form cross shaped alignments or chiasmata (diplotene, e and f). Photographs made and provided by courtesy of Professor Bernard John.

bivalents which, following dissolution of the nucleolus and nuclear membrane, become oriented around the equator of the spindle. Each bivalent is attached to the spindle at two points, the centromeres, which come to lie equidistantly either side of the equational plane. At anaphase I, the half bivalents are separated and the chiasmata resolved. The two polar groups may each form an interphase nucleus for a short time, but the second division of meiosis follows soon. It is like a mitosis mechanically in that homologous chromatids are separated after the centromeres have aligned on the equator of their spindle. But no replication of DNA occurs between the two divisions of meiosis (Fig. 5.2). Four nuclei are produced by each meiosis and each nucleus contains the haploid number of single chromatids.

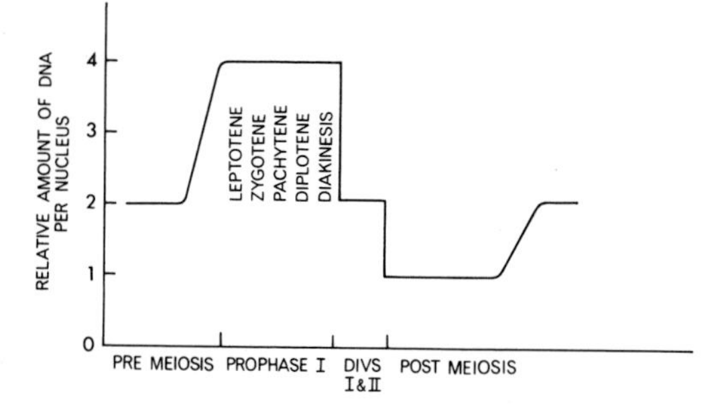

Fig. 5.2 Diagram to show the relative amounts of DNA synthesised at the pre-meiotic mitosis and throughout the stages of meiosis.

Meiosis serves to reduce the chromosome number from diploid to haploid in an orderly manner, so that one of each kind of chromosome goes into each haploid nucleus. In general, homologues are segregated from each other and the members of different homologous pairs are segregated independently. During meiosis parts of homologues are exchanged wherever the pattern of chiasma formation occurs. There are some organisms, such as the male of *Drosophila melanogaster*, and in the anthers of *Fritillaria japonica* and related species (Noda, 1975), in which no chiasmata (and no crossing over) occur. In these the homologues are held together and co-oriented for segregation by a means other than chiasmata but not understood. Very rarely a 'concealed' chiasma has been observed in the *Fritillaria* species; meiosis in the ovules shows numerous chiasmata.

The cytological evidence that chiasmata are the consequence of crossing over is strong and was analysed in detail by Darlington (1937, Ch. VII). It follows from the demonstration that sister chromatids, from the same parent chromosome, are paired on both sides of a chiasma. The proof depends upon ability to distinguish between two parent homologues. This may be achieved with respect to three kinds of properties: development, function and morphology. The distinction of development was the first to be made. In autotetraploids (Darlington, 1930), the chromosomes are paired at pachytene with exchanges of partner. If a chromosome could form one chiasma between two such exchanges of partner, a cross over must have occurred at the chiasma (Fig. 5.3a). Similar developmental distinctions occur in demonstrations using interlocking (Fig. 5.3b) or relational coiling. The distinction of function depends upon the recognition of the constant pairing properties of chromosomes. The figure of eight configuration observed in multiple interchange hybrids depends upon a chiasma which, formed between interstitial segments, must be a cross over (Fig. 5.3c). The distinction of form is the easiest to demonstrate and grasp. It is made frequently in structural hybrids, for instance in those in which two homologues differ by a segment that is inverted or by a deficiency of a segment in one of them (Fig. 5.3d; Brown and Zohary, 1955; Zohary, 1955).

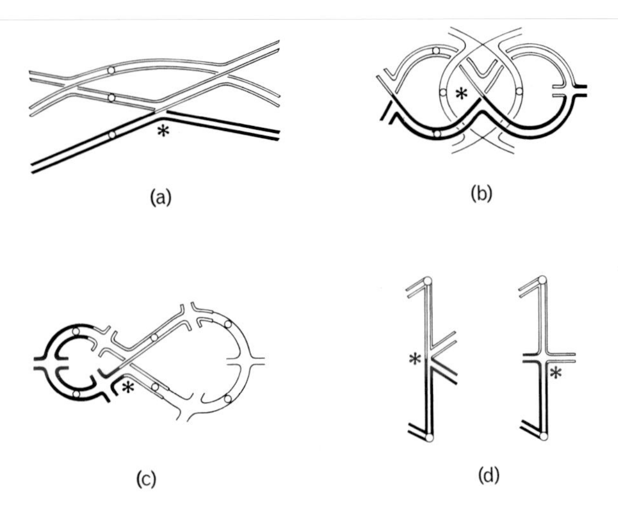

Fig. 5.3 Diagrams of meiotic chromosome configurations that prove crossing over to have occurred in the formation of the chiasmata marked with an asterisk: (a) in a trivalent, after Darlington, 1930; (b) in an interlocked bivalent, after Mather, 1933; (c) an interstitial chiasma in an interchange heterozygote, after Sansome, 1932; (d) in deficiency and inversion heterozygotes.

Stern (1931) demonstrated with *Drosophila melanogaster* and Creighton and McClintock (1931) with *Zea mays* that new chromosome arrangements were associated with recombination of genes. They used chromosomes marked by gene differences and by morphological differences at each end and were able to show that crossing over was correlated with segmental interchange between homologous chromosomes.

Direct evidence of physical exchange has been obtained by isotopic labelling of chromosomes, using ^{3}H-thymidine, and autoradiography (review: Callan, 1972). Taylor (1965) showed that exchanges between non-sister chromatids could be observed when suitably labelled chromatids of a grasshopper (*Romalea*) passed through meiosis. Although no correlation with the behaviour of gene markers could be made, the exchanges observed occurred with a frequency expected on the basis of the number of chiasmata observed. However, after certain corrections, such as sister strand crossing over, the agreement is less good. Further the distribution of points of physical exchange was random along the length of all chromosomes, whereas chiasmata were not random in distribution. Peacock (1970) has analysed the correlation in greater detail for the grasshopper *Goniaea australasiae*, and found a close relation between physical exchange and chiasma frequencies.

This correlation is analogous to the general correspondence sought between total map length and total chiasma frequency. In *Zea mays* the observed map lengths of each linkage group is less than that predicted from chiasma frequencies (Darlington, 1934), due to incomplete mapping. In *Drosophila melanogaster* where mapping is about complete, the analysis of meiosis in oocytes is virtually impossible. However, Carpenter (1975), reconstructing oocytes at pachytene from serial sections examined with an electron microscope, reports 'recombination nodules' in numbers (about 5) and in positions (on the distal euchromatin) that correlate with the expected average number (5.6) and positions of cross overs. A critical comparison was made by Beadle (1932a, b) in a *Zea-Euchlaena* hybrid between chiasma frequency and crossing over frequency in a pair of recognizable segments. Agreement was good.

Isotopic labelling and autoradiographic studies limit hypotheses of meiosis. The chromosomes are stable physical entities, even the longest chromosomes remaining essentially intact during the lengthy prophase of meiosis. A few exchanges between homologues occur, the number being very close to that predicted on the basis of chiasma frequency. Exchanges between sister chromatids are few or none. Some uncertainty is introduced by the possible effects of irradiation from incorporated tritium during the prolonged prophase; this may cause breakages capable of repair or exchange.

Three crucial problems remain: (1) the mechanism of homologous pairing at zygotene; (2) the mechanism of chiasma formation and crossing over; and (3) the basis of segregation, particularly the affinity between sister chromatids after diplotene. The second problem is better approached by genetical analyses, as outlined in Chapter 2. However evidence supplementary to that in Chapter 4, on the genetic control of events at meiosis is ample. The mechanism of pairing has been studied by electron microscopy and a particular body of genetic evidence bears on it. The basis of affinity, after pachytene, between sister chromatids is still obscure and is not treated further.

5.1 Mutations affecting meiosis and spore or gamete formation

The production of functional spores or gametes in each organism depends on the precise occurrence of numerous metabolic and biochemical processes, controlled by a large number of different genes, most of which must be present as dominants. Mutation of any to a recessive condition disturbs fertility directly or indirectly. Broadly, genetically caused abnormalities are of three kinds: (1) those which affect the differentiation of reproductive organs; (2) those which affect meiosis; and (3) those which disturb post meiotic maturation of spores or gametes. The second and third kinds cause male and female infertility despite the formation of reproductive organs. Genetic dissection of meiosis is proceeding by the collection of mutants in several different organisms. They differ in the extent to which meiosis can be investigated and whether biochemical analyses are possible. In many cases there are technical difficulties, due to sterility, in testing for allelism, a circumstance overcome in yeast by using mutants that are manifested as deficient only at an elevated temperature.

Cytologically, the mutants observed fall generally into four classes. One results in a failure to enter meiosis, exemplified by *ameiotic* in *Zea mays* (Palmer, 1971). A second, *asynaptic* (*as*), results in failure of synapsis at zygotene, and so failure of chiasma formation. In the third, *desynaptic* (*ds*), synapsis occurs at zygotene apparently normally, but subsequently there are seen to be fewer chiasmata formed, ranging to a complete failure. *Asynaptic* mutants have been observed in *Pisum sativum* (Gottschalk, 1968), while *desynaptic* mutants are reported for many species, but often called *asynaptic*, e.g. in *Zea* (Beadle, 1933). The number of different loci involved is uncertain, due to the difficulty of testing for allelism where the mutants are sterile. However, where chiasma failure is incomplete crosses between *desynaptic* mutants in *Pisum* suggest several to many loci perhaps as many as 22, with perhaps four *asynaptic* loci. In the tomato (*Lycopersicum esculentum*) six *asynaptic* loci are listed (see Clayberg *et al.*, 1966), together with 37 *male sterile* loci, presumably concerned with steps in the post meiotic development of pollen. Soost (1951) described 5 *as* loci in the tomato, while Moens (1969) showed that the reduction of chromosome pairing in three of them (*as1*, *as4* and *asb*) did not result in a reduction of the frequency of recombinants among offspring. Indeed in two of the mutants the frequency was significantly increased. The reduced fertility of *desynaptic* plants and the observed formation of defective gametes suggest that only a few gametes are viable and that the observed progeny are a selected group. However, it would be expected that the frequencies would be about equal to the controls, as in *as1*, rather than greatly raised as in *as4*. Selection does not predict elevated levels of recombination. The enigma remains, as it does in *Zea* (Rhoades, 1946) where a considerable amount of crossing over is found in diploid female gametes formed by *asynaptic* plants. The possibility that chiasmata are formed and somehow quenched needs study.

A fourth type of meiotic mutant is exemplified by *pc* (*precocious centromere division*) in the tomato (Clayberg, 1958) and is analogous to *mei-S332* in *Drosophila melanogaster* (see p. 70). It is observed that the centromeres divide prematurely at meiosis, starting at anaphase I and completed in all chromosomes by prophase II. The plants are completely pollen sterile and highly egg sterile.

5.2 Synaptinemal complex

Extension of the observations on meiosis, by electron microscopy, to higher magnifications than those obtainable with the light microscope are on the whole disappointing. However, two important points have been established. One is that at early stages of prophase all chromosomes are anchored at their ends to the nuclear membrane, all points of anchorage being in a small area of the membrane. The end of the chromosome with the nucleolar organizer is anchored to the surface of the nucleolus, itself usually in contact with the nuclear membrane. The other discovery is that in almost all eucaryotes the pairs of homologous chromosomes are locked together by a special structure, the synaptinemal complex (Moses, 1956; reviews: Moses, 1968; Westergaard and von Wettstain, 1971). The frequent spelling 'synaptonemal' is unsound. The serial sectioning of nuclei has been particularly rewarding (Moens, 1970).

In all cases, the synaptinemal complex is a ribbon shaped axis consisting of two lateral components and a central component flanked by space which is less electron dense. The space around the central component is traversed by thin filaments. The structural detail visible in different eucaryotes varies as also do the dimensions of the parts. Seen in side view the lateral components are 30–50 nm wide and are usually amorphous, but banded in fungi and some insects. In *Neotiella* the lateral component usually has alternating thick (10 nm) and thin (5 nm) bands (but sometimes two thin ones adjacent) with a repeating unit of about 20 nm, along the length of the ribbon (Figs. 5.4 and 5.5). The width of the central region is 90–120 nm with the central component 10–30 nm. The central component is amorphous usually, but a lattice in insects. The elements of the synaptinemal complex are composed of ribonucleoproteins. All components are digested by trypsin and major portions of the lateral components are digested by RNase. However, DNase has no effect on the synaptinemal complex.

The homologous chromosomes, already two chromatids at zygotene, are paired at pachytene by the synaptinemal complex and held at a distance apart of about 100 nm, irrespective of chromosome size, morphology and DNA content. The union by the complex is over the whole length of the chromosomes including the centromeres. The central component is thickened locally into distinct nodes, of which there are 0 to 3 per bivalent in *Neurospora crassa* at variable positions in a given bivalent

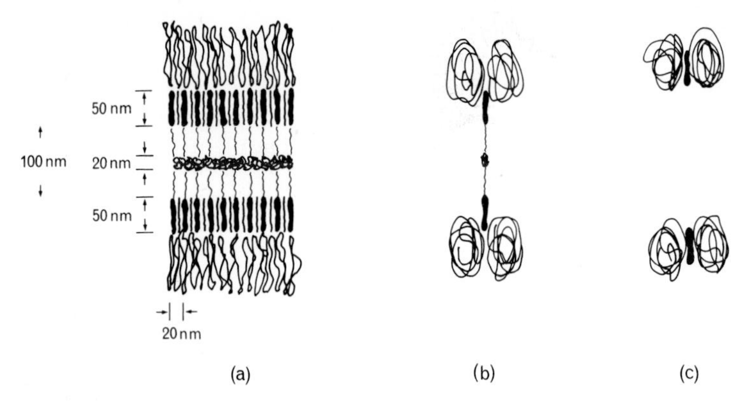

Fig. 5.4 Diagrams of the components of the synaptinemal complex in *Neotiella rutilans*, based on photographs and data of Westergaard and von Wettstein, 1970: longitudinal section (a) and transverse section (b) at pachytene; (c) leptotene in section with lateral components holding sister chromatids together.

in different cells. However, chiasmata are not observed at pachytene by either light or electron microscopy.

The lateral component appears attached to unpaired leptotene chromosomes and arises, as observed in *Neotiella*, in the groove between the two sister chromatids. The two chromatids appear single by light microscopy due to their union by the lateral component of the synaptinemal complex. Pairing at zygotene is accompanied by relocation of the lateral component to the outside of each chromosome, so that the two lateral components are in register opposite to each other and join with the central region into one synaptinemal complex.

The first stretches of complex appear at several independent places between each pair of homologues. When pairing of homologues is terminated at early diplotene and replaced by their repulsion, the synaptinemal complex is shed except where the chromosomes are held together by chiasmata. The shed parts are structurally altered into an amorphous fibrillar mass. In fungi and flowering plants both components are shed, but in mammals the lateral component returns to the original position between sister chromatids.

Several lines of evidence confirm that the synaptinemal complex functions in securing generally homologous pairing, though how it can mediate any degree of specificity is quite obscure. In triploid *Lilium tigrinum* (Moens, 1968) each of three (partially) homologous chromosomes forms a single lateral component along its entire length at leptotene. In any one region, two of the lateral components are combined into a synaptinemal complex. The attractive properties of the lateral components are satisfied by association in pairs.

In organisms which lack chiasmata, the synaptinemal complex may or

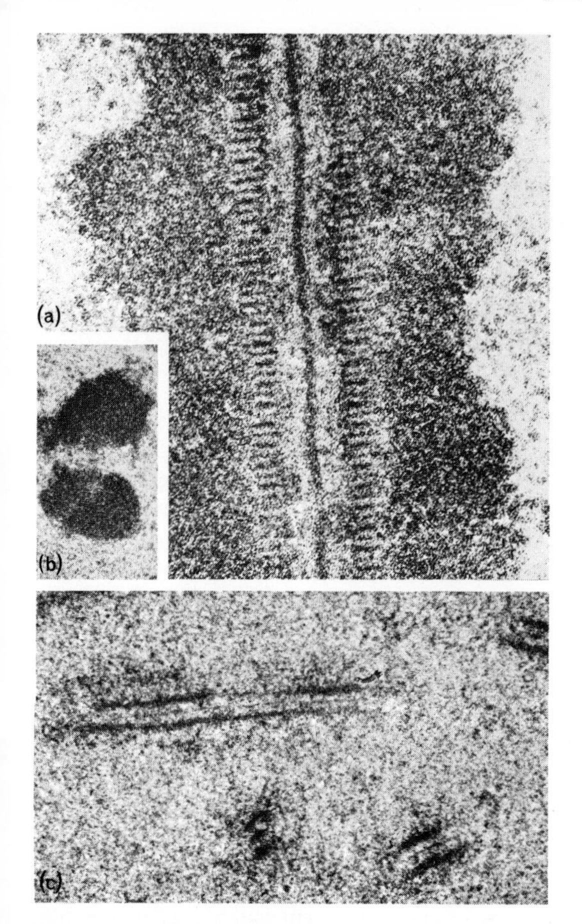

Fig. 5.5 Synaptinemal complex at pachytene in fungi: *Neotiella rutilans* (a) longitudinal (lateral) view at 90000 × and (b) transverse section at 31000 × magnification; (c) *Neurospora crassa* section of part of a nucleus at 18000 × magnification. Photographs by courtesy of Professor D. von Wettstein.

may not be present. The mantid *Bolbe* retains the complex from pachytene to metaphase-I, thus preserving chromosome pairing in the absence of chiasmata. On the other hand, there is said to be no synaptinemal complex in the male *Drosophila*. *Asynaptic* mutants lack a synaptinemal complex. The *Drosophila melanogaster c(3)G* has none. The wheat *asynaptic* has the lateral components at leptotene but no complex at pachytene. Several *desynaptic* mutants in the tomato (*as1*, *as4*, *asb*) have a synaptinemal complex at pachytene (Moens, 1969).

Recombination between homologous chromosomes requires molecular interaction. However, at pachytene of meiosis, the stage at which recombination occurs, the chromosomes are not only condensed to about 0.2% of their length, but the main masses of the homologues are about 0.1 μm apart. Hence interaction can, at the most, be between only a small fraction of the total chromosome length and must be by loops extended from the chromatids into the central component. These loops (perhaps the thin filaments that traverse the space around the central component) must be more than 0.1 μm long and therefore each comprise more than 300 nucleotide pairs. It is possible that the function of the lateral component is to activate these pairing loops and that specificity of association between homologues is achieved by the interaction of homologous pairing loops, perhaps policed to confirm true homology.

5.3 Genetic control of pairing at meiosis

Species of the genus *Triticum* may be diploid (*T. monococcum*), tetraploid (*T. turgidum*) or hexaploid (*T. aestivum*, the bread wheat) having, respectively, 14, 28 and 42 chromosomes as diploid numbers. The tetraploids and hexaploids are allopolyploids which have arisen by hybridization of differentiated species and then chromosome doubling. The species that are the sources of the sets of chromosomes in the polyploids have been discovered by study of chromosome association (by chiasmata) in hybrids with related wild species, mainly of the genus *Aegilops* if this is regarded as separate from *Triticum*. The haploid sets in the species are referred to as A in *T. monococcum*, AB in *T. turgidum* and ABD in *T. aestivum*, the A, B and D sets coming from different ancestral species. The source of set B is uncertain; the original proposal of *T. speltoides* is now discounted leaving a choice between several others (Sears, 1969; Kimber, 1974). Set D comes from *Aegilops squarrosa* (= *T. tauschii*). The seven chromosomes in each of the A, B and D sets have their homoeologues (roughly similar but not identical) in each of the other sets, as demonstrated by homology of genetic content and ability to pair and form chiasmata under special conditions. The chromosomes in *T. aestivum* are named 1-7A, 1-7B and 1-7D, the letters identifying the ancestor and the numbers the homoeologues. The tetraploid and hexaploid *Triticum* species normally form bivalents exclusively, 14 and 21 respectively, even though there is homoeology between chromosomes of the sets. This is not due to competition between identical (homologous) chromosomes and roughly similar (homoeologous) chromosomes, because removal of competition in the 'haploid' *T. aestivum* does not lead to more than rather rare pairing.

The cause of regularity of pairing was found by Riley and Chapman (1958) to be due to a gene (later called *Ph*) on chromosome 5B that has the property of suppressing homoeologous pairing or, regarded alternatively, of promoting exclusively homologous pairing. This was discovered through the systematic study of meiosis in nullisomic variants of *T. aestivum*. These plants are reasonably viable because the loss of two homologous chromosomes from one set is compensated largely by the functional genes on the four homoeologous chromosomes of the other two sets. Plants nullisomic for 5B, and therefore with 40 chromosomes, have multivalents and its 'haploid', with 20 chromosomes, shows abundant pairing. Subsequently, genes on several other chromosomes have been found to influence chromosome pairing in wheat, though to lesser extents. Mostly, the effects have been studied by removal of whole chromosomes, in nullisomics, or of parts of chromosomes, as in isochromosomes, with two like arms, or telochromosomes, with one arm only, produced by misdivision of centromeres. By these means the 5B gene was shown to be in the longer arm. Many of the studies have used hybrids between the various aneuploids of *T. aestivum* (monosomics) and a number of wild species of *Aegilops*, partly motivated by a search for the source of the 5B gene and partly by the greater viability of monosomics than nullisomics. However, there should be intense selection in allopolyploids in favour of such genes which would act to reduce irregularities of pairing and so diminish infertility. Hence *Ph* might not have existed as such in a wild ancestor.

Collectively the data, especially of Mello-Sampayo (1971), for hybrids with *Aegilops sharonensis*, and Driscoll (1972), for hybrids with *Aegilops variabilis*, show genes of minor effect on 3A, 3D and 4D and possibly very minor effects due to 1B, 3B and 4A. The presence of genes in some wild species affecting pairing in hybrids is indicated by Dover and Riley's (1972a) results. Species of *Aegilops* are either inbreeding or outbreeding (e.g. *A. speltoides*, *A. mutica*). In hybrids of inbreeders with wheat there is very little chromosome pairing unless the 5B chromosome is removed. However, hybrids of outbreeders (*A. mutica*) with wheat show high levels of pairing in the hybrids, the frequencies of chiasmata per cell falling into four classes: low (1.5), low intermediate (4.1), high intermediate (7.1) and high (12.9). These occur with frequencies which suggest two linked loci, about 23 cM apart, each with two alleles in the *Aegilops*. Omission of 5B from the wheat parent (using a 5B monosomic) leads to a 'super high' class with 17.5 chiasmata per cell. Thus, no gene that can compensate for the absence of the 5B gene has yet been found in any normal *Aegilops* chromosomes. Some of the *Aegilops mutica* plants have small supernumerary B-chromosomes. In the absence of 5B and the presence of one to four B-chromosomes*, the super high class is eliminated, showing that the B-chromosomes do carry the 5B gene that prevents homoeologous pairing (Dover and Riley, 1972b).

Several means of action of the 5B gene and of other genes with similar effects have been considered. Riley (1968) supposed that lengthening the period of meiosis through the removal of 5B might be responsible for extra pairing. However, the increased duration (of about 15%) is due to the

* The B-chromosomes are supernumerary chromosomes commonly devoid of normal genes; they must not be confused with the B-chromosome set of wheat species.

removal of the short arm of 5B. In the presence of the long arm of 5B, which carries the gene for homologous pairing, the meiotic period is normal. Others, especially Feldman (1966), have advocated a space effect. Two assumptions are made, that homologous chromosomes are not distributed at random before meiosis and that this determines homologous pairing, secondly that the 5BL gene promotes mitotic association of homologues. Observations suggest that at mitosis homologues tend to be somewhat closer together than are homoeologues, but the differences are rather small. Spatial effects due to secondary pairing, are not generally supported. Driscoll and Darvey (1970) have observed the effects of colchicine. In a strain with a 5DL isochromosome, pairing and chiasma frequency is reduced in all chromosomes except for the isochromosome, showing that colchicine acts during the leptotene-zygotene phase of pairing while the 5B gene acts after this stage.

Riley (1974) has summarized data on the action of temperature switches on chiasma frequencies which point to events in G1 of meiosis being crucial to chromosome pairing. In the absence of chromosome 5D chiasma frequencies are sharply reduced below 19°C and above 29°C. Its absence also causes failure of pairing in an isochromosome at low temperature. Evidently 5D affects a process in meiotic G1 that is necessary for pairing and is independent of the relative positions then of the potential partners.

It seems probable that genes of the kind present in 5BL have a function in monitoring pairing and eliminating pairing that is not strictly homologous. Previously, it has been noted that other chromosomes of *Triticum aestivum*, notably 3D, carry genes with properties like those of the gene in 5BL. For example, Mello-Sampayo (1971) found the following chiasma frequencies per cell in crosses of *T. aestivum* strains to *Aegilops sharonensis* (= *T. longissimum*):

euploid	× *A. sharonensis*	1.56
aneuploid (−5B)	× *A. sharonensis*	15.8
aneuploid (−3D)	× *A. sharonensis*	9.76
aneuploid (−3Dβ)	× *A. sharonensis*	10.26 (telo 3D)

The removal of 3D (and 3A which has a minor effect) is similar to the removal of 5BL. All normal genes in these chromosomes must be present for regular homologous pairing. Thus the genes on 5B and 3D and other chromosomes are complementary in their action, as though each contributes to a function which eliminates non-homologous pairing. They might each specify a different enzyme needed for the removal of erroneous pairing or, perhaps more probably, contribute to an enzyme composed of several elements all of which must be present or normal for full effect. Corrective mechanisms to eliminate non-homologous pairing may be important in all organisms. Non-homologous pairing is observed in haploids, though chiasmata are not very common in them. Association by synaptinemal complexes is very frequent in the barley haploid (Gillies, 1974). If a similar frequency of zygotene pairing were possible between homoeologues, a mechanism for terminating such pairing before chiasmata were formed would be advantageous to the organism. It is possible that the *sticky* type of gene mutation, e.g. in *Zea mays* (Beadle, 1932c), exhibits the consequence of non-homologous pairing followed by chiasma formation leading to bridges and fragments. The function of st^+ would be to remove non-homologous pairing.

5.4 Biochemistry of meiosis

Information about the biochemical events in the early phases of meiosis is scanty. The most extensive is from *Lilium* (reviews Stern and Hotta, 1973, 1974), and there is some for yeast and wheat and a few other organisms.

The premeiotic S-phase, during which the chromosomes are replicated, is considerably longer than the corresponding phase in premitosis. In *Triturus* the phase is three times as long (Callan, 1972). Nevertheless the rates of DNA synthesis, as determined by DNA fibre autoradiographs, are similar in cells destined for meiosis and those for mitosis. The extended S-phase in premeiosis appears due to fewer initiation points and so fewer replicons. A second feature of this phase of synthesis is that it is incomplete, about 0.3% of the DNA in *Lilium* being deferred in its replication until the beginning of zygotene. These interruptions of replication are considered to be confined to specific and constant regions of the chromosomes which have a distinct composition, the GC content being about 50% compared with about 40% for the chromosomes as a whole. It is not so heavy as ribosomal DNA. The evidence for incomplete replication is based on DNA–DNA hybridization methods (Hotta *et al.*, 1966) and on the incorporation of BUdR (Hotta and Stern, 1971b). The DNA synthesized during zygotene can be distinctly labelled and isolated. This Z-DNA can then be hybridized with DNA from different stages of meiosis and the relative proportions of Z-DNA per chromosome set determined. It is found that nuclei in prezygotene contain half as many Z-DNA sequences as in pachytene or later stages. Hence zygotene synthesis of DNA in *Lilium* must be a delayed replication rather than an extra round of replication. The conclusion is supported by buoyant density analysis of DNA prepared from cells exposed to BUdR, which increases the density of DNA in which it is incorporated. Such DNA prepared in cells exposed during the premeiotic S-phase shows an increase in density except for the DNA that is replicated during zygotene. The Z-DNA is identified either by radioactive labelling after the S-phase or by hybridizing fractions with a preparation of Z-DNA labelled previously. An increase in density of Z-DNA occurs only if BUdR is supplied during zygotene. The Z-DNA is more or less generally distributed among the chromosomes; ^{3}H-thymidine label introduced at zygotene is scattered. The labelled Z-DNA present at pachytene is readily separated from the rest of the DNA if prepared as an alkaline extract; it is either not covalently bonded or the covalent bonds are very labile.

During a very short period after the end of the premeiotic S-phase, not more than a quarter the period to leptotene, cells can be induced to revert to mitosis by transplant into synthetic culture medium. If reversion is induced the Z-DNA is synthesized prior to entry into mitosis. This reversion recalls the behaviour of the *ameiotic* mutant of maize (Palmer, 1971), in which presumptive meiotic cells, undergo a final mitotic division in place of the normal meiotic one. During the interval between S-phase and leptotene changes occur in chromosome organization, involving a commitment to meiosis. Presumably this is a step wise process, for various treatments applied to cells during the interval lead to abnormalities of synapsis. If an inhibitor of DNA synthesis is applied to leptotene cells, they will not proceed to zygotene and the chromosomes do not pair (Hotta *et al.*, 1966). Addition of the inhibitor during zygotene (Roth and Ito, 1967) stops the formation of more synaptinemal complex without destroying the stretches already formed.

It is speculated that Z-DNA is in the hypothetical synaptic regions of chromosomes. Chiu and Hastings (1973) have advanced this view with respect to *Chlamydomonas*, suggesting that incompletely replicated regions are the sites of crossing over. The treatment of zygotes, towards the end of the S-phase, with phenylethyl alcohol increases the frequency of recombination. The effect is attributed to an increase in the number of replicons which fail to complete replication. Alternative suggestions that the inhibitors may affect events in prezygotene or early zygotene that are essential to synapsis are no less probable. These may readily alter the places and frequency of sites of effective pairing. It can also be concluded from several lines of cytogenetic evidence that there is a determination before pachytene of the sites of crossing over, but not necessarily of crossing over itself (Maguire, 1968). The agents chiefly used are temperature shocks and colchicine treatment. Levan (1939) was the first to show that colchicine reduces chiasma frequency and he attributed the action to interference with pairing. Other work, especially on wheat (p. 98) and *Lilium*, shows that colchicine applied at leptotene disrupts pairing and that this effect extends up to early zygotene. Apparently colchicine interferes with the process of pairing. In treated pollen mother cells, the lateral elements of the synaptinemal complex are present but the complexes themselves are formed only partially or abnormally.

Many lines of evidence point to the process of crossing over occurring in the pachytene phase. The transition from zygotene to pachytene is accompanied by major changes in DNA metabolism. Once homologous chromosomes have synapsed, enzymes appear that have the functions respectively of producing single nicks in DNA strands and of repairing gaps formed (Howell and Stern, 1971). The outcome is the formation of a more or less constant number of cross overs per meiotic cell.

Two types of deoxyribonuclease have been found in *Lilium*: A with a pH optimum of 5.6 and B with a pH optimum of 5.2. The latter, deoxyribonuclease B, is found exclusively in meiotic cells, whereas A is present in all vegetative tissues and in meiotic cells of some, but not all, varieties. The deoxyribonuclease B has transient activity, beginning in zygotene, rising to a maximum in pachytene and vanishing at the end of pachytene. The nature of the nicks produced has been determined using partially purified preparations. It is an endonuclease that acts only on native DNA and makes single stranded nicks that expose 3′ phosphate and 5′ hydroxyl ends. Exonucleases are present, but their properties are not known. There is also a polynucleotide ligase, dependent on ATP, present that can seal 5′ phosphate to 3′ hydroxyl ends. The 3′ phosphate and 5′ hydroxyl ends·could be changed by a phosphatase removing the 3′ phosphate and by polynucleotide kinase phosphorylating the 3′ hyroxyl group.

Mild extraction of DNA shows the presence of a relatively higher number of relatively short strands of DNA during pachytene compared with earlier and later stages. The number of nicks per nucleus is hard to determine, but has been estimated at 10^5, vastly greater than the average of 36 cross overs per nucleus, even allowing for the need for a minimum of four nicks per cross over. Only nicks in synapsed regions are effective.

DNA is synthesized during pachytene. Autoradiographic studies in a number of organisms, e.g. wheat (Riley and Bennett, 1971), show synthesis during pachytene. However, doubts have been expressed as to whether this may not reflect the low level of repair synthesis operative in all tissues, perhaps stimulated by the radiation damage from the ^{3}H thymidine incorporated into the cells. The DNA synthesized at pachytene (P-DNA) is formed by repair replication. There is a constant proportion of it at different stages of meiosis, as shown by extraction of P-DNA and hybridization with DNA from cells at different stages. P-DNA that contains BUdR does not differ significantly in density from other DNA, even when sheared to fragments of about 5×10^5 daltons (about 1500 nucleotides). Evidently the stretches of BUdR are too short to affect buoyant density significantly. The newly synthesized DNA is scattered through the genome, but the actual size of the pieces is unknown. There is very little P-DNA formed in hybrids without chiasmata or in cells in which pairing has been blocked by colchicine.

That pachytene synthesis of DNA resembles repair replication is supported by the action of inhibitors (Hotta and Stern, 1971b). Hydroxyurea, which inhibits semi-conservative replication preferentially, is a poor inhibitor of P-DNA synthesis, but effectively inhibits the synthesis of Z-DNA and premeiotic DNA. Hydroxyurea and other selective inhibitors, e.g. 6-(p-hydroxyphenylazo)-uracil, affect S-phase and Z-DNA synthesis in one way and radiation induced and P-DNA synthesis in another.

There appears to be selectivity in the sites at which pachytene repair of DNA occurs, suggesting that events occur at preferred sites. According to Smyth and Stern (1973) the P-DNA contains a relatively higher proportion of reiterated sequences than does nuclear DNA in general. Studies of the reannealing of DNA after melting shows that P-DNA does so rather quickly, indicating the general similarity of the sequences, scattered through the chromosomes, that exhibit synthesis of DNA at pachytene. The synthesis of P-DNA usually involves regions with a class of sequences that has a limited number of varieties in the genome. The lengths of these common sequences are not known, but the results suggest that if nicking of DNA marks the beginning of the crossing over process, the nicking occurs at preferred places. It is proposed that these are in or close to loci containing one of several special varieties of repeated sequences in the genome. They may correspond to the *con* loci deduced in *Neurospora crassa*.

It should be made clear that similar analyses of meiosis in wheat (Riley and Bennett, 1971; Flavell and Walker, 1973) do not show the same patterns of DNA metabolism found in *Lilium*. Synthesis of DNA occurs throughout meiosis, but the synthesis at zygotene is not distinctive in base ratio, nor does inhibition of synthesis prevent chromosome pairing.

Turning from DNA metabolism to evidence for special proteins at meiosis, the general labelling of proteins has not disclosed any that may be regarded as of particular significance. However, a possibly significant exception concerns proteins associated with the nuclear membrane at meiosis (Hecht and Stern, 1971). The Z-DNA is associated with newly synthesized protein in a lipoprotein complex and is selectively released in this association by gentle homogenization of the nuclei. On the contrary P-DNA is not associated with protein. The association of Z-DNA with protein is transient and is gone by pachytene.

Separation of the proteins of the nuclear membranes (Hotta and Stern, 1971a) shows the presence of a denser fraction (s = 1.2–1.22) in addition to those of the outer (1.16–1.18) or inner (1.18–1.2) membranes also found in somatic nuclei (s = density in g/ml). The special fraction contains two specific proteins one of which binds to DNA and the other to colchicine. The DNA that binds to protein is similar in several respects to the gene *32* protein produced in bacteria infected with T4 and essential to the recombination of the phage. The lily protein facilitates the renaturation of single stranded DNA and is not specific with respect to the source of the DNA, acting equally well on T7 as on lily DNA. It is present only in meiotic cells, appears during leptotene, reaching a maximum at zygotene, begins to disappear at mid-pachytene and is quite gone at the end of pachytene. Temporally, it is associated with the period of chromosome synapsis, which perhaps it facilitates.

The other protein, that binds colchicine, is apparently not exclusive to meiosis, though there is a considerable increase in the quantity of it during the meiotic prophase. It is present in all three fractions of the nuclear membrane. It appears to be similar to somatic tubulin except for its smaller molecular weight (about 10^5 daltons compared with 1.2×10^5) and in not being precipitated by vinblastine. The relation of this protein to the period of sensitivity to colchicine is close and it may be that this protein is also concerned in synapsis.

6

Recombination in bacteria

'The hereditary units can hardly differ much in nature from the lowest and simplest organism'.

Charles Darwin, *The Variation of Animals and Plants under Domestication*

6.1 Mechanisms of genetic transfer and recombination

The transfer of genetic information from one bacterium to another is mediated in several different ways, different especially in the size of the DNA transferred and in the mechanism of transfer. They include conjugation, transduction, transformation and episomal carriage. Following transfer, genetic recombination between recipient and donor usually occurs by a mechanism apparently similar to that of gene conversion in eucaryotes. The genetic maps (Figs. 6.1 and 6.2) incorporate features referred to particularly in this chapter (*E. coli* map, Taylor and Trotter, 1972).

Conjugation (Curtiss, 1969; Susman, 1970) involves the association of two cells and the transfer of part or all of the chromosome from one cell to the other. Conjugation of cells requires the presence in one cell (the male) of a conjugal fertility factor (F) and its absence from the other (the female, F^-). Transfer of the bacterial chromosome requires that F be integrated into the chromosome. The F factor determines the formation, on the cell surface, of pili which couple cells together and may serve as conduits through which chromosomal material is conveyed from male to female. The F factor may be autonomous (F) or integrated (Hfr), the latter allowing the transfer of chromosomal material from the Hfr male to the F^- female. The place of integration of F is one of a number of specific loci in the bacterial chromosome, the preferred locus depending on the strain. It has been shown that the *E. coli* chromosome contains, at various positions, several copies of certain sequences of DNA called 'insertion sequences' (IS) (Saedler and Heiss, 1973). There are eight of IS1 (about 800 base pairs long) and five of IS2 (about 1400 base pairs). The F factor possesses both IS1 and IS2. Recombination between one of these sequences and its homologue in *E. coli* leads to insertion of F so that it is flanked by a repeated sequence (say IS1) at each end. The Hfr positions (Fig. 6.1) are where IS loci are, the direction of orientation of F being determined by the direction of the IS sequence in the *E. coli* chromosome. The IS sequences also have the property of translocation from their standard positions to other parts of the chromosome, frequently causing mutation. In this regard, they act as controlling elements of gene activity (Saedler *et al.*, 1974). This illegitimate recombination, not requiring extensive homology of DNA, is not controlled by the same enzymes that operate regular recombination.

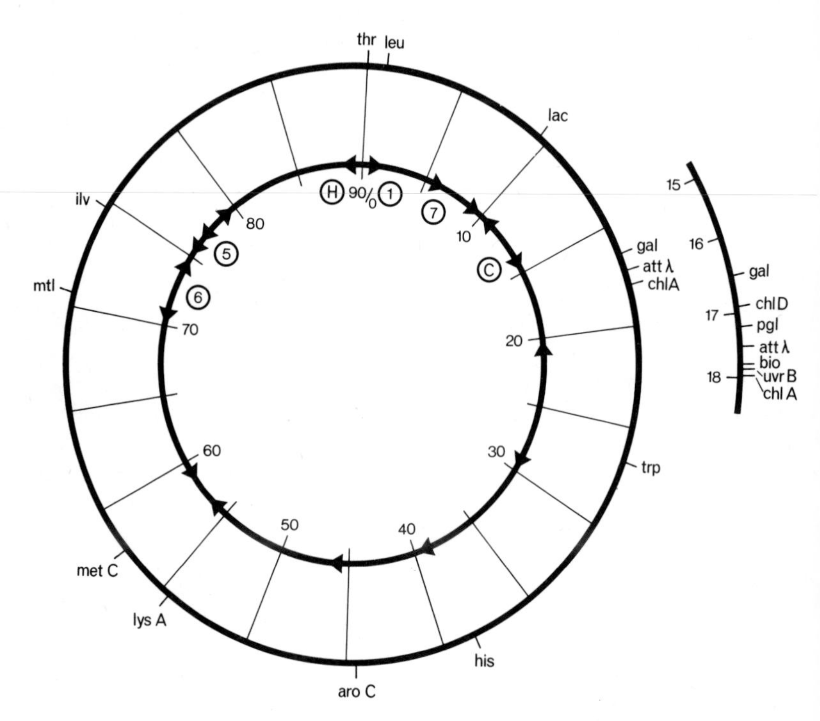

Fig. 6.1 Linkage map of *E. coli*. The inner circle bears a time scale, in minutes, based upon the results of interrupted conjugation. It is graduated in five minute intervals beginning arbitrarily with zero at the *thr* locus. The inner circle also bears arrows which mark the origin and direction of transfer for several *Hfr* strains; some of these are lettered or numbered in a ring. The outer circle bears the symbols of a number of gene loci, one region around the attachment site for lambda phage being expanded on an outer arc. The gene symbols and the function of the normal genes are as follows: *thr*, threonine synthesis; *leu*, leucine synthesis; *lac*, lactose operon; *gal*, galactose operon; *chlD*, nitrate-chlorate reductase; *pgl*, 6-phosphogluconolactonase; *attλ*, integration site for phage lambda; *bio*, biotin operon; *uvrB*, repair of radiation damage to DNA caused by ultraviolet light; *chlA*, nitrate-chlorate reductase; *trp*, tryptophan operon; *his*, histidine operon; *aroC*, chorismic acid synthetase; *lysA*, diaminopimelic acid decarboxylase; *metC*, cystathionase; *mtl*, mannitol; *ilv*, isoleucine-valine operon.

Transfer of the chromosome begins at the locus of F integration and proceeds in a given direction in an orderly sequence. Transfer of the whole chromosome is rare, so the genes closer to the origin are transferred much more frequently than those more distant. A part of the integrated F factor is transferred at the leading end of the Hfr chromosome, but the rest is left at the distal terminus. Consequently, the transfer of donor properties which depends on the transfer of the whole F factor, with the whole chromosome, is rare.

It appears that only one of the two strands of the donor chromosome is transferred to the recipient and it is believed that transfer is accompanied by replication of the donor chromosome, one of the two replicas appearing in the recipient as synthesis proceeds. The transferred DNA fragment is known to consist of one donor DNA strand and one newly synthesized strand, the latter possibly made in the recipient cell. It has been demonstrated (Ihler and Rupp, 1969) that only one of the two strands can be recovered from recipients after conjugation with labelled lysogenic donors (p. 119). Indeed, opposite strands of the prophage (p. 119) were found in the recipients when donors which transfer their chromosome in opposite directions were used. Evidently a specific strand of the F factor is broken and transfer starts at this site, commencing with a free 5′ terminus on the donor strand.

The mechanism by which exchange occurs between donor and recipient chromosome is apparently similar to that in T4 phage in that the first detectable product is a joint molecule (Oppenheim and Riley, 1966; Cooper *et al.*, 1971). In the joint molecule, the parental DNAs are joined by non-covalent bonds. Later, the parental DNAs become joined by covalent bonds (Oppenheim and Riley, 1967; Cooper *et al.*, 1971). Using improved methods, Paul and Riley (1974) have obtained better yields of joint molecules from $F^- \times$ Hfr and shown that they have a molecular weight of about 4×10^7. The parental Hfr segments ranged from 3×10^6 to 11×10^6 with a mode of about 9×10^6 molecular weight. Parental F^-DNA, parental Hfr DNA and new DNA synthesized during mating each made up between a quarter and a half of the total content of the joint molecules. Gaps in the joint molecules, capable of infilling under the influence of phage T4 polymerases, were usually no greater than 450 nucleotides long. Recipients that were mutant, either *recA* or *recB recC*, were able to form joint molecules, though *recB recC* showed a reduced amount. However a mutant F^-, deficient for the region homologous to the proximal end of the Hfr, formed no joint molecules.

Apparently, complete integration usually does not occur in the immediate recipient of the transferred piece of chromosome. Anderson (1958) and Lederberg (1957) made pedigree analyses of the progeny of single zygotes, isolating them by micromanipulation, often over several generations. Anderson observed that most recombinants appeared late, after the third generation and up to the ninth. Moreover, more than one type of recombinant, inheriting some of the same donor genes, frequently issued not only from the same zygote but in different generations of the same line. No progeny of the Hfr donor ever displayed factors derived from the F^- parent. Many progeny of the F^- cell were inviable. It was suggested that the injected fragment of the Hfr cell was replicated and passed to a number of progeny, so that several occurrences of recombination could follow one donation. Lederberg (1957) found that the majority (52) of 75 pedigrees he studied contained only one kind of recombinant while others had up to four different ones.

Episomal transfer (Campbell, 1960, 1962, 1969) is dependent upon sex factors and other plasmids that are capable of transfer from cell to cell and may at the same time transfer bits of the host's chromosome. Plasmids and episomes differ from phages in not being encapsulated and so being restricted in dispersal. The mating system of *Escherichia coli* comprises two types of male strains, F^+ and Hfr, which have different properties with respect to genetic transfer and bacterial phenotype. Both produce filamentous pili on the cell surface and these enable them to conjugate with receptive female (F^-) cells. The pili also confer susceptibility to a number of spherical RNA viruses and some filamentous DNA viruses to which F^- cells are resistant. The F^+ character is readily transmitted to F^- cells independently of the bacterial chromosome, thus resembling a virus infection. Treatment of F^+ strains with acridine orange in low concentration causes loss of the F factor, converting the cells to the F^- state. Such 'cured' cells can be converted back to F^+ by conjugation with an F^+ strain; they gain simultaneously the ability to produce sex pili and to be susceptible to phages specific to donor bacteria. These and other properties show that, in F^+ cells, the F factor is autonomous, existing as a separate cytoplasmic element distinct from the bacterial chromosome. The properties of F, particularly with respect to the polarity and linear nature of chromosome and F transfer, were suggested (Jacob and Monod, 1963) to be explained by F as well as the chromosome (Cairns, 1963) being circular and by F being inserted into the bacterial chromosome as λ phage is. Transfer replication starts by initiation in the F factor, presumably at a specific site, and proceeds in a polarized fashion around the chromosome.

Hfr bacteria differ fundamentally in several ways. They arise with a low frequency from F^+ bacteria and promote conjugation in which there is a high frequency of transfer of parts of the bacterial chromosome from donor to recipient F^-. But the recipient stays F^-, a necessary part of the F factor being transferred only at the distal end of the bacterial chromosome. There are several different kinds of Hfr strain, different in origin and in direction of transfer. The Hfr character is not destroyed by acridine orange. The change from F^+ to Hfr is the consequence of a single reciprocal recombination event, which results in the insertion of F at one of several possible loci in the *E. coli* chromosome (see p. 104). The F^+ and Hfr chromosomes are both circular and it was inferred that the F factor is also circular (now confirmed by electron microscopy) and is integrated by a mechanism similar to that by which λ phage is inserted into the *E. coli* chromosome (see p. 131).

Hfr strains are able to generate altered sex factors, F′, which each incorporate a piece of the bacterial chromosome. They first appeared as intermediate donor strains (I) that transmitted the same oriented sequence of bacterial chromosome as the original Hfr, but with only about a tenth of the frequency. The F′ could be transferred readily to recipient F^- cells, which were then convertible to secondary intermediate donors (I′) with properties similar to the original. Treatment of the I strain with acridine orange converted it to an F^-. Reinfection of this same strain with a normal F^+ sex factor converted it back to an intermediate donor because a part of F remained in the B chromosome. The interpretation is that the intermediate donors have an F′ factor, consisting of a normal F factor (or the substantial functional part of it) plus a segment of the bacterial chromosome. It is presumed that this arose from a rare recombination between a part of the F factor integrated into the Hfr chromosome and a neighbouring region of the chromosome. The origin would be analogous to the formation of λdg factors (λ defective, galactose transducing—also called $\lambda dgal$) that are able to transduce *gal* (see p. 131). Because the segment of bacterial chromosome carried by F′ is homologous with the corresponding segment of the chromosome of a recipient, the frequency of insertion and release of F′ would be much greater than for F, resulting in a rapid alternation of Hfr and F′ states. The result would be a population of bacteria (the I′ secondary donors) which can transfer both the autonomous F′ factor and the Hfr chromosome at high frequencies, as observed in the I′ donor strains.

The F′ factor contributed very much to the analysis of the *lac* operon system, among many others, since partial diploids may be constructed readily. Heterozygotes (often called heterogenotes) like the partial diploid $Bz/F'z^+$ are unstable for the z^+ character, segregating z progeny with a frequency of 10^{-3} per generation. They also generate inverse heterogenotes ($Bz^+/F'z$) as well as the homogenotes ($Bz/F'z$). These occurrences reflect recombination between the bacterial chromosome and its homologous segment in the F′ factor. Such recombination is often, though not always, reciprocal. Herman (1965) constructed a strain heterozygous at two loci of the *lac* operon, namely $Bz^+y/F'zy^+$. Only recombinants would be able to ferment lactose and therefore be either Bz^+y^+ or $F'z^+y^+$. Of the recombinants, eleven were found to have $F'z^+y^+$, by testing by transfer to a suitable F^- tester (Bzy). After treating the originals with acridine orange to remove the episome, the B genotype could be determined. Seven of the eleven were found to have the reciprocal Bzy genotype, the remaining four having non reciprocal genotypes. It would be interesting to know what proportion are not reciprocal if the sites of difference were closer together and in the same gene locus.

Evidence of predominantly reciprocal recombination between well spaced genes in λ prophage has been obtained by Meselson (1968), using a strain in which the prophage is present both in the chromosome (B) and in the episome (F). He studied recombinants arising from B*c mi h*/F+++ diploids by determining the types of prophages present in a large number of unselected cells. The intervals *c mi* and *mi h* are similar, showing 2.64 and 3.17% recombination respectively. Of 5734 cells analysed, 5136 had parental combinations (+++, *cmih*), 112 had a pair of reciprocally related combinations (36+*mi h*, *c*++; 68++*h*, *c mi*+; 8+*mi*+, *c*+*h*), 162 had only one type, while 325 had a pair that were not reciprocally related. These data do not in themselves provide evidence for or against reciprocal recombination. However, the demonstration is made by comparing recombination in the two intervals. For example, if there were no true reciprocity of recombination in these intervals, the combination *c*++ would be found as often with ++*h* as with its reciprocal +*mi h*. However, the observation of only 8 *c*++/++*h* and 8 *c mi*+/+*mi h* shows these to be many fewer than the 36 *c*++/+*mi h* and 68 *c mi*+/++*h* reciprocals.

Further information about the origin of F′ shows they all not only contain some bacterial DNA but also some or all of the fertility factor's (F) DNA. They are produced by insertion of F into the bacterial chromosome by a reciprocal recombination to produce an Hfr. F′ DNA is produced by an aberrant excision to make a circular DNA with at least the essential genes of F plus some bacterial DNA. The F is about 6×10^7 daltons, compared with the bacterial 2.5×10^9 daltons. Each F′ is distinctive though there are preferred sites for excision as well as preferred sites on the F for its insertion into the *E. coli* chromosome. The structure of different F′s has been studied by Sharp *et al.* (1972) by making heteroduplexes and examining them by electron microscopy.

Transduction (review: Ozeki and Ikeda, 1968) is the transfer of a small part of a bacterial chromosome from donor to recipient by the agency of a bacteriophage. It may be general or specific. In general transduction, e.g. by P22 in *Salmonella typhimurium* or by P1 in *Escherichia coli*, almost any part of the donor's chromosome may be transferred. In contrast, the λ phage in *E. coli* K12 is capable of transducing only genes (such as those at the *gal* and *bio* loci) that are close either side of the attachment site specific for the λ prophage.

Two theories of general transduction have been advanced: (1) that, as in special transduction, fragments of bacterial chromosome are incorporated into the phage chromosome, replacing a part of it, to yield defective phages, like F′ factors; (2) that a piece of bacterial DNA is by chance enclosed within the head protein envelope in place of the phage chromosome. In the case of P1 and P22 the evidence is that both situations are possible, though the latter is the commoner. Defective transducing phages, similar to λdg, contain bacterial genetic material from the neighbourhood of the attachment site for the phage, but not apparently from elsewhere. It has been observed that about a quarter of

P1 phages are smaller than normal (about 70% of the diameter). The small phages are defective, each carrying a part of the phage chromosome such that, in multiple infections, they can complement one another. A proportion of the small phages are transducing, but carry smaller pieces of bacterial chromosome than normal, as shown by the reduced probability of co-transduction of neighbouring genes.

Transformation (review: Hotchkiss and Gabor, 1970) is the process by which DNA extracted from donor cells is incorporated into recipient cells; sometimes also called transfection when virus DNA is transferred. It was first described for *Diplococcus pneumoniae* (Pneumococcus) and formed the system by which DNA was first shown to be the genetic material (Avery *et al.*, 1944; McCarty *et al.*, 1946). It occurs in many other bacteria, notably *Haemophilus influenzae* and *Bacillus subtilis*. Each system shows variability with respect to: (1) the proportion of cells transformed, due to the requirement for a special physiological state of the recipient, called competence; (2) the size and nature of the DNA molecules that are effective; (3) the fate of the DNA molecules during the eclipse period between absorption into the recipient cell and integration into its chromosome. Competence may reflect the presence of receptors or of enzymes necessary for passage through the cell wall or the absence of nucleolytic enzymes that may destroy the DNA. Very little is known of the receptive sites for attachment of DNA to the cell, or of the mechanisms of transport. Fortunately these gaps in knowledge are not important to the study of recombination itself.

Native DNA is the most efficient in transformation; denaturated DNA separated into single strands has very little activity (Kent and Hotchkiss, 1964). The size of the pieces of DNA is important. Cato and Guild (1968) showed that fragments of 0.3×10^6 to 0.8×10^6 molecular weight are tightly bound to pneumococcal cells, and transform actively. Although smaller fragments are taken up by cells, their transforming activity diminishes rapidly with decreasing size. Synapsis, presumably between specific pairing regions, requires a minimum critical size, corresponding to about the length of a gene or a cluster of genes (operon). Double stranded DNA is necessary immediately prior to synapsis. Integration into the recipient chromosomes occurs fairly quickly, in 10 to 15 minutes. During the eclipse period a complex of donor and recipient material is detectable; it has a molecular weight of about 750 000. Presumably it is a heteroduplex or a more complex association.

In *Diplococcus pneumoniae*, Fox and Allen (1964) showed that either chain of the donor DNA formed a hybrid with a complementary chain of the recipient's DNA and without any substantial amount of synthesis of DNA occurring. The hybrid extends over 2000–3000 nucleotide pairs, which is shorter than the average length of the DNA when it first enters. The donor material is covalently linked into the DNA of the recipient bacteria. These conclusions are based on experiments in which donor DNA was labelled with heavy atoms (^2H and ^{15}N) and radioactive atoms

(^{32}P), the donor and recipient also each having a different mutation. The DNA of transformed bacteria was analysed by density gradient separation both before and after denaturation (into single strands) and cutting into fragments by sonic vibration. In *B. subtilis* similar experiments by Bodmer and Ganesan (1964) and Bodmer (1966) are consistent with the above.

Thus only one strand of the donor DNA finally participates in integration. At an early stage the transformed cells are often heterozygous. Genetic recombination occurs in a manner analogous to allelic recombination in eucaryotes. Presumably integration is accomplished by nicking one recipient strand at a site where a gap is opened to allow one donor strand to pair with the other intact recipient strand. Bodmer (1965) showed that the available gap is more often near the growing point of the chromosome. Pairing would migrate until the complete donor strand is inserted. At the same time the other donor strand and the displaced part of the recipient strand would be degraded by exonuclease action. Thus a heteroduplex segment would be formed and be subject to excision and repair to yield pure transformed clones by gene conversion or form mixed clones by replication and segregation.

There are divergent reports, reminiscent of those on fungi, about different responses of different kinds of mutant, differing in the efficiency of their integration or appearance in recipient's progeny. This would correspond to differences in probability of transformation (conversion) to the donor state. In *Diplococcus*, Lacks (1966) distinguished four classes of mutants, deficient in the amylomaltase gene, by integration efficiencies in the ratio of 1:4:10:19. He suggested tentatively, on the basis of the mutagens used to produce the mutations, that these might correspond with the four possible kinds of base substitution, namely: (1) AT for GC or GC for AT; (2) AT for TA; (3) GC for CG; (4) AT for CG or GC for TA. In heteroduplexes there would be eight possible kinds of mispairing of bases, as discussed in Chapter 8; it is not obvious why they should group into four probability classes. Indeed Ephrussi-Taylor and Gray (1966) found only two classes, with a ratio of efficiencies of 1:10 among mutants resistant to aminopterin. Lacks found that small deficiencies or duplications integrated efficiently, but that larger ones showed decreasing efficiency. In double mutants the efficiency of integration is not additive, but is that of the less efficient one if the two are linked closely. However, using three linked factors, Kent and Hotchkiss (1964) found that all combinations of donor characters occurred with nearly equal frequency and concluded that the molecular events were complex. It might appear that several exchanges occurred over the length of the donor DNA fragment.

Lacks attributes differences to the effect of the nucleotide difference on the probability of integration while Ephrussi-Taylor and Gray attribute the effect to greater probability of excision of a donor nucleotide mispaired, accompanied by extensive excision of neighbouring material.

Bresler, Kreneva and Kushev (1968, 1971) made analyses of clones, of *B. subtilis*, by transferring individual transformed cells to non-selective medium before they could replicate. A number of different factors were found to show 20 to 80% of pure clones, the rest mixed, depending upon the particular mutant used. The occurrence of mixed clones indicated that the original molecule had remained hybrid, whereas the pure clones showed there had been a correction of mismatching prior to the first replication. The timing of the events would correspond to post meiotic and meiotic segregation in eucaryotes. They concluded that correction of mispairing is frequent; moreover closely linked mutants were often corrected together, showing co-conversion.

Darlington and Bodmer's (1968) results also show that excision and repair of individual mismatched nucleotides is unusual. They exposed cells carrying four linked mutations (*aro2*, *try2*, *his2* and *tyr1*) to DNA from prototrophic cells, using DNA at a limiting concentration. Some 2000 transformed cells, with the two outer donor genes, were examined for the two inner genes. Only 15% of the cells were not also transformed for these inner factors. They conclude also that exchanges between the selected genes, with either of the intermediate ones, must be rare.

Another approach to the analysis of conversion is that of Spatz and Trautner (1970) who constructed heteroduplex donor DNAs from SPPI phage and used these to infect recipient cells of *Bacillus subtilis*. The heteroduplexes were made by annealing complementary strands of wild type with those from 21 different plaque mutants of a phage. The separation was accomplished by melting the DNA, annealing it with ribosomal RNA or poly IG (a copolymer of inosinic and guanylic acids) and separating heavy and light strands on a caesium chloride density gradient. One strand of the DNA hybridizes selectively with the ribonucleotide. The technique is similar to the one used by Doerfler and Hogness (1968) to determine which strand of λ phage carried the information for a given *sus* gene. The progeny observed are phages rather than bacteria. The heteroduplexes constructed from the wild type and mutant phages of various mutagenic origins show a proportion, ranging to about 50% of infections, in which no correction occurs (Table 6.1). In these the two strands participate equally as shown by the distribution of the two types of progeny in single bursts. In the rest of the infections there is nearly always inequality favouring either the light strand or the heavy strand in the case of mutants induced by hydroxylamine or nitrosoguanidine. The behaviour of double mutants, either in coupling or repulsion is not readily predictable.

Spatz and Trautner attempted to test the specificity of conversion by inducing mutations with hydroxylamine, which produces almost exclusively C to T changes. Thus the strand containing the altered pyrimidine would be known, the wild being CG and the mutant TA, and all mispaired bases would be CA and TG. On the assumption that conversion involves purine excision, the preferred strand should be the

Table 6.1 Progeny composition from transfection of *Bacillus subtilis* with heteroduplex DNA prepared by annealing one or other strand of a mutant phage with the complementary strand of wild type (+) or another mutant. Mutants were produced by treatment of intact phage: 515, 510 and 161 with hydroxylamine; 111 with nitrosoguanidine; 141 with acridine. The heavy (H) strand was selectively hybridized with a ribonucleotide to allow separation from the light (L) strand. Data of Spatz and Trautner, 1970.

Genotype of		Percentage of bursts with			Preferred strand
heavy strand	light strand	wild and mutant	wild only	mutant only	
+	515	45	53	2	H
515	+	51	12	37	H
+	510	43	17	40	L
510	+	49	37	14	L
+	161	23	69	8	H
161	+	15	5	80	H
+	111	26	20	54	L
111	+	32	41	27	L
+	141	32	42	26	H
141	+	43	48	10	L
+ 161	111 +	27	20	53	
111 +	+ 161	28	53	19	
+ +	111 161	30	28	42	L
111 161	+ +	11	5	84	H

one with the C to T change. However, in only five out of nine mutants obtained in this way was the preferred strand the one carrying the pyrimidine change. Evidently, the specificity of conversion is dependent on a more complex mechanism and not on the nature of the mismatched bases alone. The mutant 141 is apparently a frame shift mutant which has a small deficiency; correction is preferentially to wild type whichever strand is mutant in the heteroduplex.

It was found that there was a correlation between the frequency of transfections (a form of infection using naked DNA and a competent host) in which there was no conversion and the map position of the mutant concerned. The frequency declined from a relative high in mutant *515* to a relative low in mutant *NOH3* (a mutant produced by treating a heavy strand with nitrous acid) these two mutants spanning about three-quarters of the map. Evidently replication starts proximally to the *515* site and proceeds towards (and beyond) *NOH3*, so that more distal mispairs would have more time to convert than the more proximal ones. Conversion is not affected by any of a number of genetic or physiological deficiencies of the host, e.g. recombination deficiency, reactivation deficiency or heavy irradiation of the host cell with ultraviolet light prior to transfection.

6.2 Genetic control of recombination in bacteria

The knowledge of genes concerned in recombination and the enzymic reactions they specify is more comprehensive for *E. coli* than for any other organism except T4 phage. So also is information about DNA metabolism in general. Therefore consideration of what activities are concerned in recombination and what are not can be more fully defined. Several recent reviews, especially by Clark (1971, 1973, 1974), Radding (1973) and Davern (1971), cover the topic more fully than is possible in this chapter. Mutants defective in genetic recombination are found directly by testing clones of cells, that have been subjected to mutagenesis, for strains which show fewer recombinants when tested with an Hfr donor (Clark and Margulies, 1965) or a transducing phage or DNA from a suitable donor (Beattie and Setlow, 1971). The donor is unlikely to contribute the normal allele of any mutant one affecting recombination if only a small piece of the donor's chromosome is transferred. The putative mutants must also be shown to accept donor DNA. Probably as many as ten genes are known, genetic defects in which result in reduction in the frequency of recombination (*recA* to *recL*). Also there are suppressors (*rac*, *sbcA* and *sbcB*) of the mutants *recB* and *recC*.

Many mutants defective in recombination also show increased sensitivity to ultraviolet light and ionizing radiations, or inability for prophages to be induced or for the cells to become lysogenic (see pages 128–131). However, many mutants exceptionally sensitive to ultraviolet light do not show any effects on recombination. The prime effect of ultraviolet irradiation is to cause the formation of dimers of adjacent pyrimidine bases in the same DNA chain. These interfere with the replication of DNA and if not removed lead to death of the cell. Removal is effected in any of two or three ways. Light in the visible spectrum provides the energy for a photoreactivation enzyme to split dimers to which it binds. Mutation of the gene for this enzyme leads to loss of the capacity for photoreactivation. Another system, dark repair, results in the excision of the pyrimidine dimers and usually a few neighbouring bases so creating a gap in a DNA chain. The gap is repaired by the action of a polymerase and ligase. At least three genes ($uvrA^+$, B^+ and C^+) are known to be concerned with the process of excision. Mutation of these genes appears to have no effect on recombination. Mutation of some other genes leads to sensitivity to radiations and also to other effects. Thus, at *uvrE* there is an effect on mutation rate. At *lon*, radiation induces mutant cells to grow into filaments; the *lon* mutant is present in *E. coli* strain B. At *uvrF* (equals *recF*), recombination is affected in certain genotypes.

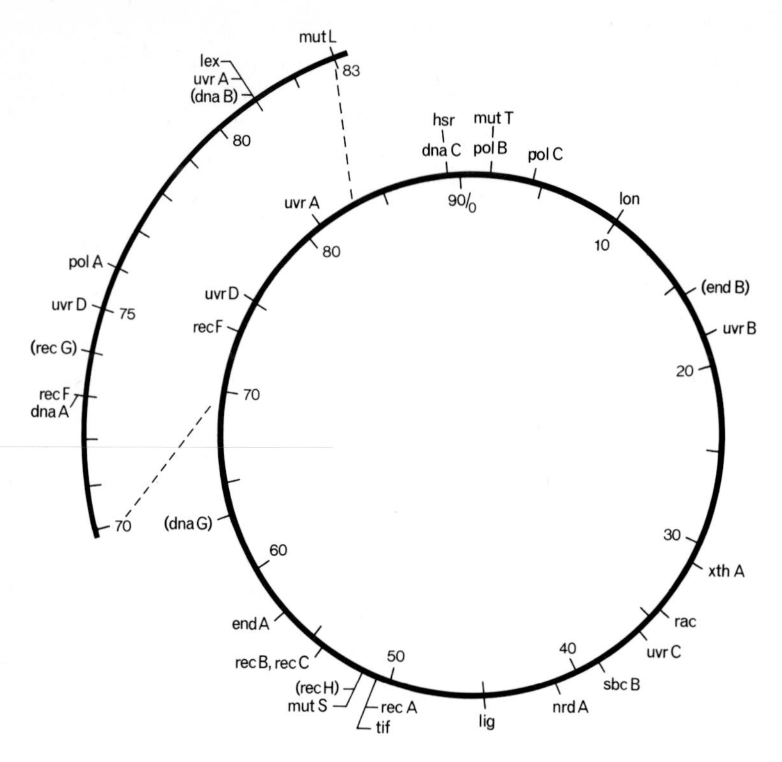

Fig. 6.2 Linkage map of *E. coli* showing the location of genes concerned with the metabolism of DNA, especially recombination. The following table lists the loci, their location on the time map (those in brackets are approximate) and the function of the normal genes.

Locus	Time slot	Function affected by mutation	Locus	Time slot	Function affected by mutation
Genes involved in recombination			sbcA		suppressor of *recB*, *recC*
recA	51	recombination; UV sensitivity	sbcB	38	suppressor of *recB*, *recC*; exonuclease I
recB	54	recombination; UV sensitivity	rac	34	recombination activation
			lig	45	DNA ligase
recC	54	recombination; UV sensitivity	endA	57	endonuclease I, specific to DNA
recF	73	recombination UV sensitivity	endB	(16)	endonuclease I, specific to DNA
recG	(74)	recombination	hsr	89	host specificity; endonuclease R
recH	(52)	recombination			

Locus	Time slot	Function affected by mutation	Locus	Time slot	Function affected by mutation
Genes involved in other aspects of DNA metabolism					
polA	76	DNA polymerase I	xthA	31	exonuclease III
polB	2	DNA polymerase II	uvrA	81	UV sensitivity
polC	4	DNA polymerase III	uvrB	18	UV sensitivity
dnaA	73	initiation of DNA synthesis	uvrC	36	UV sensitivity
			uvrD	75	UV sensitivity
dnaB	(81)	DNA synthesis	lex	81	UV and X-ray sensitivity
dnaC	89	initiation of DNA synthesis	tif	51	mimicry of UV irradiation
dnaG	(63)	regulates initiation of synthesis of Okazaki fragments	lon	10	radiation sensitivity; long cells
nrdA	42	*dnaF*; ribonucleoside diphosphate reductase, unit B1	mutL	83	high mutability
			mutS	52	high mutability
			mutT	2	high mutability; specifically AT→CG
nrdB	42	ribonucleoside diphosphate reductase, unit B2			

Several genes are known to be concerned with the synthesis of DNA, particularly replication of the chromosome. Conditional mutants, causing a defect in a function at a higher temperature or under other special conditions, have served to identify up to ten distinct genes. The genes $polA^+$, $polB^+$, and $polC^+$ (is same locus as was called *dnaE*; *polC* is used in map) are the structural genes respectively of DNA polymerase I, II and III and are potentially involved in repair processes. The gene lig^+ is the structural gene for DNA ligase.

The $recA^+$ *gene*

Mutations in this gene block recombination completely or nearly so (Clark, 1971). They also lead to high sensitivity to ultraviolet light (Clark and Margulies, 1965) and ionizing radiations (Howard-Flanders and Theriot, 1966), enhanced breakdown of DNA following irradiation (Clark *et al.*, 1966) and insensitivity to the mutagenic action of ultraviolet light (Witkin, 1969). Further, lambda prophage cannot be induced if the host is *recA*. The *recA* mutants also block the action of *tif*, a temperature conditional mutant that causes induction of λ prophage (Castellazzi, George and Buttin, 1972) at the non-permissive temperature. The *recA* mutants also promote cell division and prevent filament formation in strains, like *lon*, which would show this phenotype following irradiation with UV. The prime function of $recA^+$, whose loss leads to these pleiotropic effects, is unknown. Clearly, the function must be central to the process of recombination, for no other gene is known to replace it. However, it is not clear whether $recA^+$ functions directly or indirectly, as a regulator of the synthesis of enzymes involved in recombination.

The recB⁺ and recC⁺ genes

The two loci occur very close together but are distinct since *recB* and *recC* mutants complement one another in repulsion phase (Willetts and Mount, 1969). Further studies (Storm *et al.*, 1971) have disclosed an additional four genes in the cluster. Mutation in any component of this cluster leads to the loss of the enzymic activities associated with exonuclease V (Goldmark and Linn, 1972). It is an exonuclease, dependent on ATP, that degrades double or single stranded DNA to oligonucleotides. The action on double stranded DNA occurs simultaneously in the 3′ to 5′ and 5′ to 3′ directions in double stranded DNA and in either direction in single stranded DNA. It is also an endonuclease, cutting single strands of native DNA under the stimulus of ATP. Also, it degrades ATP in the presence of double stranded DNA irrespective of whether the DNA is digested. In mutants, recombination is reduced to between a fifth and a three hundredth of normal when tested by conjugation or transduction. Furthermore, mutant cells are more sensitive to irradiation with ultraviolet light and X-rays and tend to die spontaneously more readily.

The sbcA⁺ and sbcB⁺ genes

These are detected by the ability of their mutants to suppress the effects of *recB* and *recC* mutants on recombination. In the *sbcB* mutants exonuclease I is inactivated (Kushner *et al.*, 1972), while in *sbcA* mutants exonuclease VIII is derepressed (Barbour and Clark, 1970). The presumed structural gene specifying exonuclease VIII is designated *recE⁺*, but is not yet identified genetically.

Other rec⁺ genes

About seven other *rec* genes are known. These include the unmapped mutant *recD* (*rec34* of van de Putte *et al.*, 1966) together with *recG* and *recH* (Storm *et al.*, 1971) which result in recombination deficiency in otherwise wild type strains. The mutant genes *recF*, *recJ* and *recK* (and *recL*) result in recombination deficiency in the *recB recC sbcB* strain (Horii and Clark, 1973).

Interactions

The gene cluster, of which *recB⁺* and *recC⁺* are best known, specifying exonuclease V appears to provide functions necessary for recombination. These are an endonuclease activity, cutting DNA strands, and an exonuclease activity digesting DNA strands away. These might appear to be events relatively early in the process of recombination, presumably between synapsis (not identifiable in *E. coli*) and repair processes in which DNA is reconstructed, perhaps preceding the formation of joint molecules, in which homologous complementary strands of DNA are associated by hydrogen bonds between bases. Nevertheless, there appear to be two other ways by which, as it were, the function of exonuclease V in recombination can be replaced if it is defective. One is by the inactivation of exonuclease I in *sbcB* mutants (Table 6.2, line 8), the other by the activation of exonuclease VIII in *sbcA* mutants (Table 6.2, line 7). Clark (1974) interprets the interactions as a consequence of independent pathways, concluding that there is a *recE* pathway opened by the presence of exonuclease VIII and a *recF* pathway opened by the absence of exonuclease I, as well as the usual *recB recC* pathway. However, the two supplementary paths are detectable only if exonuclease V is inactive. All three mechanisms appear to produce DNA with gaps and so allow new associations during the process of repairs. From the point of view of the morphology of recombination, the term 'pathway' may be misleading. Rather it may be that an event, perhaps early in recombination, may be brought about in three ways.

Table 6.2 Interactions between mutant *rec* genes in *E. coli*. Based on data of Clark, 1974 (Table 2).

Locus	*recA*	*recB/recC*	*recF*	*sbcA*	*sbcB*	
Product of normal gene	Unknown	Exo V	?	Exo I	Exo VIII Repressor	Relative level of recombination
	+	+	+	+	+	100
	m	+	+	+	+	0.01
	m	m	+	+	+	0.002
	m	m	+	m	+	<0.001
	m	m	+	+	m	<0.001
	+	m	+	+	+	2
	+	m	+	m	+	100
	+	m	+	+	m	67
	+	m	m	+	+	0.02
	+	+	m	+	+	100
	+	m	m	m	+	1
	+	m	m	+	m	0.01

Other endonucleases are known (Radding, 1973; Lehman, 1967; Richardson, 1969) in *E. coli*, including endonuclease I, that breaks both strands, and endonuclease II, that breaks single strands of native DNA. Mutants deficient in these are apparently unknown. Other endonucleases are the restriction endonucleases, that break DNA so that self joining ends are produced. Two are known in *E.*

coli, namely RI and RII. They break down (restrict) any susceptible foreign DNA that may infect the bacterium. They each attack a specific sequence in the DNA, making staggered cuts in sequences

that are symmetrical, such as

$$\begin{matrix} & \downarrow & & & & \\ G & A & A & T & T & C \\ C & T & T & A & A & G \\ & & & & \uparrow & \end{matrix}$$

which is attacked by RI. These are interesting in that they recognize specific sequences. Endonucleases of corresponding specificity may be involved in at least some situations such as the site specific recombination between *E. coli* and λ phage or between the F plasmid and *E. coli*.

7

Recombination in bacteriophages

'. . . *having nothing, and yet possessing all things*'.

II Corinthians vi, 10.

Bacteriophages are viruses that grow in bacteria. Two categories, virulent and temperate, are distinguished commonly. Virulent phages invariably multiply rapidly in the bacteria they infect and destroy them completely by lysis. Temperate phages have the alternative that following infection, they may either follow the lytic cycle or establish an integrated symbiotic relation with their host cells, multiplying in step with the division of the host. The choice of which path is taken is dependent very largely on the physiological state of the bacteria; the probability of one or the other happening is strongly affected by environmental conditions. Bacteria which carry temperate phage integrated with them are said to be lysogenic, since various agents can induce termination of the dormant state of the phage and cause entry on the lytic phase. The state of the phage in the lysogenic bacterium is called prophage. Phages in the state where they are growing actively in bacteria are called vegetative phage, the mature phage outside bacteria being infective phage. Phages in the lytic cycle may undergo recombination with one another. They enter and leave the prophage state by recombination involving a special locus in the chromosome of the host bacterium and a locus in the phage. The distinction between virulent and temperate phage is convenient rather than profound. Moreover, virulent mutants of temperate phages arise frequently. Recombination will be considered in examples of each type.

Phages T2 and T4 of Escherichia coli

These are closely related virulent phages, T2 being the one in which viral recombination was discovered (Delbrück and Bailey, 1946; Hershey, 1946). They have a life cycle starting with the infective stage which exists outside the host and by which they may be dispersed. They have a head and a tail, with a complex of tail fibres, by which the phages become attached to the host cell wall at specific sites. The contents of the phage, principally a linear molecule of DNA, are then discharged into the host leaving the empty head and tail outside. The DNA now inside is the vegetative phage and is the form in which it reproduces and also undergoes recombination (reviews: Broker and Lehman, 1975; Mosig, 1970a), detectable if more than one kind of closely related phage is present in a host cell. For a time no mature phage can be detected in the host if it is broken open artificially. After a while a few can be recovered by inducing lysis prematurely with cyanide or chloroform, the number increasing with time after infection until, after a latent period, the host bursts or lyses spontaneously liberating a number of infective phages. The latent period is characteristic of the phage and of the physiological conditions; so also is the burst size, the number of phages liberated at the time of lysis. Single bursts, comprising the progeny grown in one host, can be studied by diluting infected bacteria during the latent period and distributing them to a large number of individual tubes, many of which will contain either one bacterium or none.

The mature infective phage T4 contains one linear molecule of double stranded DNA. Its molecular weight is about 130×10^6 daltons, equivalent to about 2×10^5 nucleotide pairs. The length of the DNA molecule, observed in the electron microscope, is 53 ± 1.5 μm, a value agreeing with that (52 μm) observed by Cairns (1961) using tritium autoradiography. The DNA is unlike that of most other organisms in having hydroxymethyl cytosine (HMC) in place of cytosine. Moreover, in T4 all of the HMC is glucosylated at the hydroxymethyl group, 70% in α-glycosidic linkage and 30% in β-glycosidic linkage. In T2, 25% of the HMC is not glucosylated and 5% is diglucosylated. The glucose in T2 is linked to the hydroxymethyl group by a β-glycosidic linkage; the second glucose, when present, is also linked to the first by a β-glycosidic linkage. It is usual for the mature phage to contain rather more than a complete set of the genetic material, there being a duplication of about 1–3% of the whole. The duplication is terminal, one end repeating the genes at the other end of the linear molecule. The particular parts duplicated are quite random, as though chopped out of a continuous string, abcd..... xyzabcd.....xyzabcd.....xyzabcd....., with the cuts at random, though not always providing a complete set of genes.

The genetic markers used in studying phages comprise those with characters determining the size and appearance of the plaques (lysed areas) formed in a lawn of bacteria and the host range of the phages. Additionally, a large range of mutants of T4 have been obtained that are either unable to grow in the host at higher temperatures (Edgar and Lielausis, 1964) or in a particular host due to *amber* mutations (Edgar, Denhardt and Epstein, 1964). These have allowed the identification and mapping of the genes concerned in various functions in the growth and reproduction of the phage, since the mutants can be maintained in permissive conditions and their phenotypes examined in restrictive conditions (Epstein *et al.*, 1963; Edgar and Wood, 1966; Edgar and Lielausis, 1968). The genetic map (Fig. 7.1) is fairly comprehensive

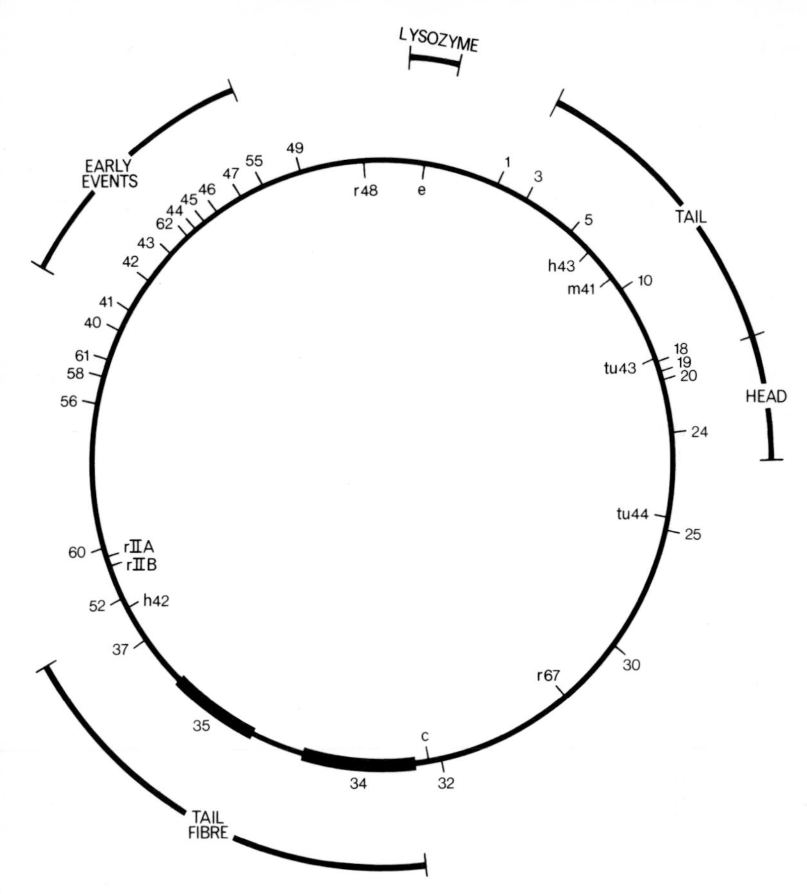

Fig. 7.1 Bacteriophage T4 genetic map, based on Wood (1974) and other data. The numbered genes on the outside of the circle are defined by conditional lethal (temperature sensitive or amber) mutants. Numbers 1 to 49 were originally defined and placed in numerical order, clockwise from a point near locus *e* (lysozyme); numbers 50 to 62, discovered later, are interpolated. Some other genes are marked on the inside of the circle; these include *host range* (*h* and *r*), *minute plaque* (*m*), *turbid plaque* (*tu*) and *clear plaque* (*c*) loci used in classical experiments on T4. Loci concerned with particular functions are grouped in special parts of the chromosome, thus: loci 3 to 18 for the tail, loci 19 to 24 for the head, loci 34 to 37 for the tail fibres and loci 41 to 47 for early events in the growth of the phage.

(Stahl, Edgar and Steinberg, 1964; Mosig, 1968; Wood *et al.*, 1968; Wood, 1974; Russell, 1974).

Recombination analysis (Mosig, 1970a) involves the infection of bacteria simultaneously with two, or more, different phages, differing in mutant characters. Hershey and Rotman (1949) found no correlation between the numbers of reciprocal recombinants present among the progeny from single infected bacteria and concluded that recombination was not reciprocal. Some progeny with factors from all three parental strains were found after simultaneous infection with three different mutants of T2. Visconti and Delbrück (1953) showed that this could be explained if successive opportunities of recombination occurred during the replication of the phage in the host cell. They postulated that, in the vegetative pool, the phage chromosomes pair and exchange DNA randomly with respect to choice of partner. Thus the yield from one infected bacterium is a population arising from several successive generations in which mating and recombination could have occurred. Hence to observe the mechanism involved in a single recombination it is necessary to adopt special methods. These include inducing premature lysis so that the consequences of single events can be observed or by using mutant strains which, though able to undergo some stages of recombination, are unable to complete the process or to replicate.

The genetic map of T4 is linear, unbranched, of finite length, but without ends. It is circular (Streisinger *et al.*, 1964). The first crosses that established this configuration involved genetic differences at three loci. When the sequence of three loci, *a*, *b* and *c*, was determined it was found that they were linked in that order. When *c*, *d* and *e* were examined, again the order was *c d e* and so on in sequence to *x y z*. When *y*, *z* and *a* were tested it was found that the order was *y z a*, with *y* and *a* tightly linked. These results can be represented only by a closed curve, a circle for convenience.

Hershey and Chase (1952) found that for each gene of T2 tested about 2% of the progeny following mixed infection by mutant and normal phage were heterozygous for the factor. Two explanations were suggested (Levinthal, 1954): (1) that a segment of duplex DNA of each genotype was present in the phage; (2) that there was an overlap of single DNA chains covering the locus, one chain mutant and the other normal. Both kinds of heterozygosity have been demonstrated (Fig. 7.2). The former may be called redundancy, the latter heteroduplex, heterozygotes for convenience of reference. Both are important to recombination, but in different ways.

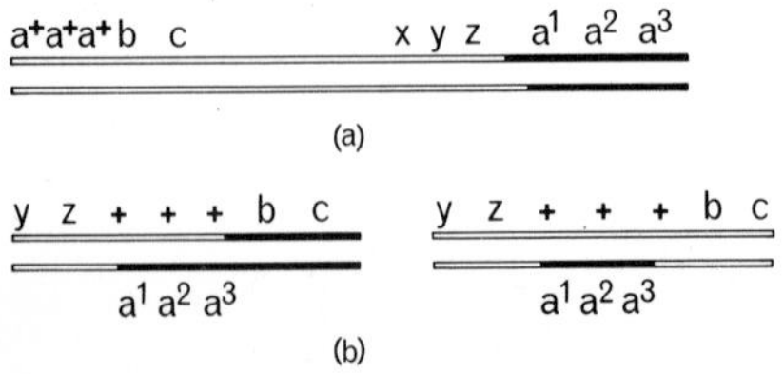

Fig. 7.2 Bacteriophage T4 illustrating the two types of heterozygote encountered at a locus; (a) terminal redundancy; (b) two forms of heteroduplex.

Doermann and Boehner (1963) used crosses involving several closely linked *r*II mutants and a plaque morphology factor less closely linked to *r*II. They found short heteroduplex heterozygotes with both ends in the marked *r*II region. Many had only one *r*II mutant and in these the homozygous markers each side of the heteroduplex region were parental. The two alleles of heterozygous sites appeared in equal frequency among the progeny. By contrast, in redundancy heterozygotes, the regions were long, with only one end in the *r*II region. The homozygous markers outside were in recombinant configuration. The markers appeared in increasingly unequal ratios, said to be 'polarized', the closer they were to the end of the heterozygous region.

Womack (1963) crossed strains with five mutations in *r*IIA with strains with two mutations in *r*IIB and absorbed the progeny singly to *E. coli* K(λ), which is restrictive to *r*II$^+$ recombinants and to terminal redundancy heterozygotes with complete *r*IIA and *r*IIB genes, so that any mutant sites are covered. Most heterozygotes showed polarized segregation, but nearly 2% of the progeny from a burst were again heterozygous for at least one *r*II mutant site. Most of these daughter heterozygotes were of the heteroduplex type, though usually for only one or a very few *r*II sites.

By experiments of these kinds, it has become evident that each phage chromosome has some three to ten heterozygous regions, mostly heteroduplex. The basis is that the average length of each heteroduplex region is no more than 0.1 to 0.3% of the total map length, while the total frequency for each genetic marker is about 1%.

Interference in phage crosses, involving the correlation of exchanges, may be of two kinds. However, positive interference, involving the interference of one exchange with another is not found except in special conditions. Negative interference, as associated with conversion in eucaryotes, is usual, with double cross overs in three point crosses being about 1.6 times as many as expected for weakly linked factors. The coincidence increases considerably for closely linked factors, such as *r*II mutants. This high negative interference results in recombination frequencies not being additive over short genetic distances. The duplicate ends of infecting chromosomes are in different map intervals, so their increased recombination frequencies contribute to heterogeneity and so to positive correlation (negative interference) in recombination in mass lysates.

Genetic recombination between T4 phages may also be studied by physical means, labelling one or both input phages in different ways with tritium, radioactive phosphorus, heavy isotopes of carbon or nitrogen or with analogues of bases. The progeny may be separated then by differences in buoyant density and classified for genetic markers and chemical differences. The phage DNA molecules are broken and rejoined in new conformations in recombination. The extent of breakage, measured physically, agrees reasonably with the extent of exchange measured genetically. Genetically there is an average of twenty exchanges per genome and these tend to be clustered to some extent. Isotope transfer experiments show that progeny receiving any parental DNA label have only a fraction, from less than 5% to more than 50% (Kahn, 1964), the average being 5–10%. The size distribution of the pieces is bimodal, about 50% of the material being in relatively large pieces, the rest being in segments less than one per cent of the chromosome. There is a stepwise formation of recombinant DNA molecules. Some recombinant segments of vegetative T4 DNA have the components of different parental origin held together by hydrogen bonds, between paired bases. These are thought to be intermediates in molecular recombination because they appear before covalently linked recombinant molecules are formed.

Some 66 different gene loci, mostly identified by conditional temperature sensitive (*ts*) or amber (*am*) mutants, have been mapped (Fig. 7.1). Some of these are concerned with the metabolism of DNA and have been examined for their possible involvement in recombination. Table 7.1 lists all the known loci which may affect recombination, together with the known gene products and effect on sensitivity to ultraviolet light (Baldy, 1968). With mutants sensitive to elevated temperatures (*ts*), the method has been to measure recombination at a series of temperatures (Bernstein,

Table 7.1 Bacteriophage T4, gene loci concerned with the metabolism of its DNA, especially recombination. The loci are arranged in order clockwise in the standard map (Fig. 7.1). Symbols: + = increase, − = decrease, 0 = none, N = normal.

Gene locus	Normal gene product or function	Effect of mutational deficiency	
		Recombination	Sensitivity to ultraviolet light
30 (*lig*)	Polynucleotide ligase	+	+
32		0	
56	Deoxycytidine triphosphatase	+	N
58			
61	DNA replication	N/+	
40			
41	DNA replication	N/+	
42	Deoxycytidine hydroxymethylase	+	
43 (*pol*)	DNA polymerase	+	+
62	DNA replication	N/+	
44			
45		N	N
46	DNA nuclease	−	
47	DNA nuclease	−	+
55	DNA replication	N	
49			

1967, 1968). With *amber* mutants, Berger *et al.* (1969) compared mutants and controls at the same time after infection, while Kutter and Wiberg (1968) and Wiberg (1966) used different times in order to compare bursts of the same size. Some disagreement on whether mutation at a particular locus alters recombination is consequential on the different procedures.

Gene 30

Replication of DNA in phage proceeds in the 5′ to 3′ direction; on one strand the growth is towards the fork and on the other away from the fork by formation of a series of small fragments (Okazaki fragments; Okazaki *et al.*, 1968a, b) subsequently joined by a ligase. Mutants (*am*) of gene 30 do not induce the formation of polynucleotide ligase when grown in the non-permissive host (Fareed and Richardson, 1967; Hosoda, 1967). Also temperature sensitive mutants induce a ligase that is more heat labile than the normal. The ligase catalyses the esterification of an internal 3′-OH group with the adjacent 5′ phosphomonoester; in the reaction ATP is converted to AMP with the release of inorganic diphosphate. The reaction works only if no nucleotides are missing at the junction. Newman and Hanawalt (1968) reported that small (possibly Okazaki) fragments are accumulated by a temperature sensitive mutant at the restrictive temperature. It is presumed that the stimulation of recombination is a consequence of the accumulation of more breaks than normal. Certainly, *gene 30* mutants growing under restrictive conditions are able to form covalently joined physically recombinant molecules (Richardson *et al.*, 1968), presumably with the aid of host ligase. *E. coli* has a ligase that requires DPN (not ATP) as a co-factor (Pauling and Hamm, 1968).

Gene 32

The product of the normal gene has been purified and the reaction catalysed by it is apparently unwinding of the DNA (Alberts, 1970; Alberts and Frey, 1970) to form single stranded regions. The infecting DNA of mutants is replicated only once and no joint molecules are formed (Tomizawa *et al.*, 1966; Kozinski and Felgenhauer, 1967). The joint molecules, composed of two components derived from parental DNA molecules and joined by hydrogen bands, are intermediates in the formation of physically recombinant molecules.

Gene 56 and gene 42

The normal *gene 56* specifies deoxycytidine triphosphatase, while normal *gene 42* specifies deoxycytidine hydroxymethylase. These enzymes catalyse steps in the synthesis of 5-hydroxymethyl deoxycytidylate (dHMP). The limited availability of dHMP might lead to misincorporation of other bases into DNA, increasing the likelihood of breakage and rejoining (Bernstein, 1967).

Gene 43

The normal gene specifies a DNA polymerase. A defect might introduce structural errors in DNA, so stimulating recombination. Specifically, the defective polymerase might tend to leave nicks in the chains, increasing the probability of exchange occurring.

Genes 46 and 47

Mutants do not break down bacterial DNA (Wiberg, 1966), implying that the normal genes specify deoxyribonucleases. Whereas *gene 30* amber mutants cause failure of accumulation of phage DNA in a restrictive host (Epstein *et al.*, 1963; Warner and Hobbs, 1967), the double mutants of *gene 30* with either *gene 46* or *gene 47* do accumulate phage DNA, showing that the normal *genes 46* and *47* break down damaged phage DNA. Recombination is greatly reduced in mutants of *genes 46* and *47*. Three possibilities are: (1) An exonuclease might be necessary to pare back one of the strands at a broken end to expose the complementary strands. Such exposure might be necessary for base pairing during the rejoining process. (2) An exonuclease might prune away excess lengths of single strands after the two parts have associated by base pairing. (3) An endonuclease might catalyse the formation of nicks and the reduced ability to do this would inhibit breakage and rejoining.

Genes 41, 45, 55 and 61

Mutants in these genes apparently synthesize little or no DNA in their non-permissive host, but there is no effect on recombination. Berger *et al.* (1969) who compared controls and *amber* mutants at the same time after infection, not at equal burst sizes, reported an increase in recombination; this may be a consequence of selection. The use of *ts* mutants shows no effect with increasing temperature in the cases examined.

Other genes

Harm (1964) showed that a mutation (*x*) which increases sensitivity to ultraviolet light also decreases recombination in the *r*II locus by a factor of 3.5. No gene has been recognized as responsible for the endonuclease causing single strand breaks to increase in frequency after T4 infection.

Structure of replicating and recombining DNA

Vegetative phage DNA is not easily extracted from the host bacteria but, when removed, appears by electron microscopy to consist of a tangle of threads emanating from a core. Some at least of the loops are much longer than mature phage. The threads contain breaks of single strands and also regions that are single stranded. It will be recalled that the genetic map is circular, but there is no clear evidence of closed curves in vegetative phage, though it would be easy to generate a ring from the injected DNA that has terminal redundancies. It would be necessary only to cut back the ends of chains to expose sequences that were homologous and would allow base pairing. In support of such a possibility is the observation that replication apparently begins at a fixed point, mapping near to *gene 42* (Mosig, 1970b). Incomplete chromosomes start replication only if they possess this region. It is not known positively whether a replication fork proceeds in each direction. If it did not, circularity would be essential to completion of the process. Then a continuous replication, by the 'rolling circle' mechanism (Fig. 7.3) would be a possibility (Gilbert and Dressler, 1968). However, the weight of evidence is against the

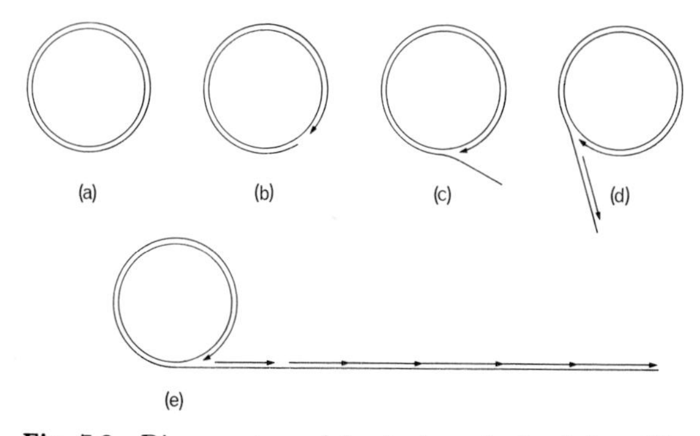

Fig. 7.3 Diagram to explain the hypothesis of the rolling circle mechanism of replication of phage T4. In (a) the phage is circular and in (b) an endonuclease has nicked one strand. In (c) the 3′ end (marked by an arrow head) of the nicked strand has begun to grow and this continues in (d), where replication in the 5′ to 3′ direction has also begun on the displaced strand. In (e) there is a later stage where a polymeric strand is begun. Such polymers may also be formed by end to end pairing of shorter strands (see Figs 8.12 and 8.13).

rolling circle as the prevailing mechanism (Doermann, 1973) and in favour of bidirectional replication from a number of genetically determined starting points. Vegetative phage appears to consist of repetitions of complete genomes arranged in tandem as though polymerized. Whether the successive genomes are covalently bonded through the sugar phosphate chains or hydrogen bonded between bases in joints is not clear. As will be shown later, there are also branches in the complex. In maturation, pieces are cut from the polymers to provide roughly enough to fill the head of each mature phage. This is usually 1.01 to 1.03 of the complete genome, so that there is 1–3% duplication.

Further knowledge of the events leading to recombination have flowed from comparative studies of the biophysics and morphology, as seen by electron microscopy, of the molecules formed by mutant phages. Broker and Lehman (1971) have used amber mutants at the *gene 30* (*lig*), *gene 32* and *gene 43* (*pol*) loci, either singly or in combinations. DNA was extracted from infected cells by a method designed to reduce fragmentation of it. The most significant new observation was of branches in hybrid molecules, seen first of all in mixed infections (crosses) at high multiplicities of the *pol* and *lig* mutants in the non-permissive host *E. coli* BBI (Anraku *et al.*, 1969). There is a positive correlation of hybrid density (using isotopic label on one parent phage) with the presence of branches. The hybrid molecules, which appear also in single mutant infections, are variable in length and in the number and arrangement of branches. Generally, more than half of the branched molecules had a single branch in the form of a fork (Y-shaped), less often a cross shape (though these are possibly artefacts), while still others had several branches. Some of these were H-shaped with the cross piece 200–2000 nucleotide pairs long. The time course of events is as follows. By eight minutes after infection, about half of the molecules have one to several nicks of phosphodiester bonds, with an average of one per 56 μm length of single strand. At 15 minutes, there is an average of eleven nicks per strand or one per 2.5 μm length of double stranded DNA. By 30 minutes the frequency reaches its maximum of about 14 nicks per strand. The mutants *pol* and *gene 32* accumulate fewer nicks.

In the *pol lig* double mutants the branches appear after fifteen minutes rising to a maximum of about 26% of all molecules at 50 minutes; at that time about 11% of the molecules are branched in a multiple fashion. The *gene 32* mutant reduces the frequency of branches by an order of magnitude. The mutant *pol* on its own accumulates fewer branched molecules than does *pol lig* as well as fewer nicks. The appearance of branches is dependent on multiple infection of the host. None appear in cells infected with only one phage, because no growth occurs in the non-permissive host.

7.2 Lambda bacteriophage

λ phage (review: Hershey, 1971) is a temperate phage able to integrate into the *E. coli* chromosome at a site between the *gal* and *chlA* loci. It illustrates a number of other differences from T4. Its DNA is a linear molecule of definite length less than a quarter the size of T4, having cohesive single stranded ends by which it can form a ring or aggregates by concatenation. The two ends are different, the right end being able to join a left end but not another right end. The joined ends dissociate if heated above a critical temperature, which depends upon the ionic strength of the medium. The cohesive ends are actually short complementary single stranded polynucleotides, one at each end of the double stranded molecule. Infection is dependent upon the presence of the free cohesive ends. Their sequences have been determined by synthesizing, with the aid of DNA polymerase, a labelled oligonucleotide complementary to each of the single stranded ends. Each proves to be a sequence of twelve nucleotides, predominantly G and C, that is complementary to the other (Fig. 7.4).

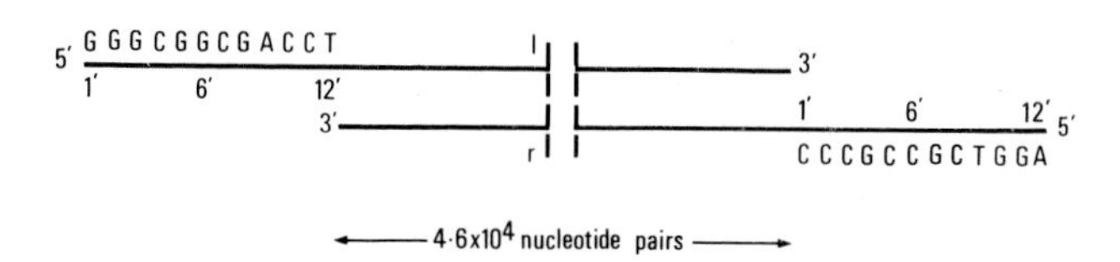

Fig. 7.4 Lambda bacteriophage, illustrating the molecular structure of its DNA, particularly of the free ends of the infective phage.

The genetic maps of lambda phage are illustrated in Fig. 7.7. The map of the lytic phage is circular, while that of the prophage is an arc of the larger circle of the *E. coli* chromosome. As a result of integration some genes, such as *J* and *int*, that are fairly close in the circular map of the lytic phage are widely separated in the prophage.

In the lytic cycle the linear DNA, after injection, rapidly assumes a circular structure 17 μm long. Although united initially by hydrogen bonds, the nicks are soon sealed by a DNA ligase, already present in the bacterium, to form 'closed circles'. It reproduces in this form, commencing at an initiation site located in the right half of the molecule near to the right end of the genetic map and proceeding at first to the left. When bacteria are infected with λ phage having isotopically dense DNA, most of the closed circles quickly become isotopically hybrid. Later, closed circles without any atoms from the infecting phage are found. Replication of this sort is semiconservative. By 15 minutes after infection about twenty circles are present in each bacterium. 'Nicked circles' with a break (or perhaps several breaks) in one chain are also found; they appear to be intermediates in the replication of closed circles. In electron micrographs of molecules in the first round of replication, most of those that can be followed completely have three loops joined at two points of branching (Fig. 7.5a). Two of the loops (a and a′) are alike in length and are the replicated parts. The two segments a and b together equal the characteristic length of the λ DNA molecule. Expressed as a fraction of this length the a and a′ segments range from nearly zero to nearly unity. Other configurations observed could be regarded as derived from the common structure by breakage.

The two branch points could be regarded respectively as the site of initiation and the growing point in replication, but in fact both are growing points. It is possible to analyse the structure by observing its molecular denaturation. Exposure of DNA to high pH or high temperature for a certain time in the presence of 10% formaldehyde leads to partial denaturation that can be seen as loops in electron micrographs. The order of appearance of loops and their size depends upon the severity of treatment and is thought to depend on the adenine and thymine content. In replicating phage DNA the two branch points move around the ring in opposite directions. Observations of the positions of the two forks point to an origin near to 0.80 on the molecular map and the termini converge at position 0.30, the antipode of the origin.

After completion of the first round of replication, most molecules are circular but may be supercoiled. Similar molecules are observed during later rounds of replication but, in addition, polymers, either circular or linear and consisting of two or three units, are frequently found. They may return to monomers by breakage and dissolution at the terminus locus.

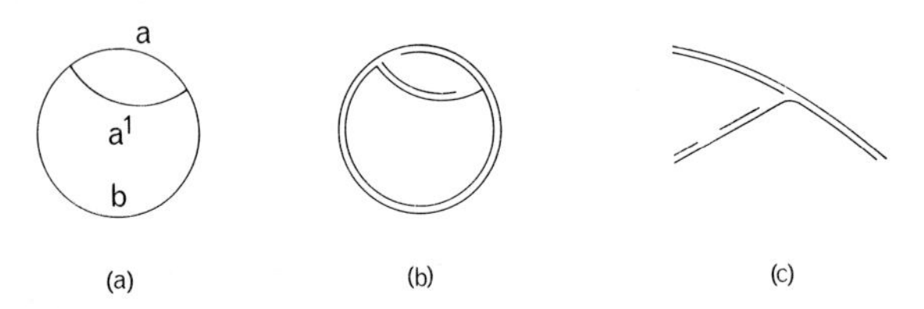

Fig. 7.5 Lambda bacteriophage, illustrating the replication of the DNA of the circular lytic phage. In (a) the forks at the ends of the aa′ loop are the growing points shown in more detail in (b) and, for one fork, in (c).

A considerable proportion of replicating DNA circles have short single stranded segments, many of them at both branch points (Fig. 7.5b). They are always located on one of the segments that has just replicated, never on an unreplicated segment. In those cases where a single strand is present at both branch points they are on different segments. In some cases the single strand is interrupted by a short double strand (Fig. 7.5c). The DNA polymerase, believed to be concerned in DNA replication, adds nucleotides to an existing chain only in the 5′ to 3′ direction. Hence at a replicating fork, only one strand (the 3′ to 5′ strand) is used as a template (Fig. 7.6). Near the replicating fork therefore there would be a segment of the complementary strand that had not replicated. It has been supposed that the gap is filled by growing short segments in the 5′ to 3′ direction and then attaching these by ligase action to the previously replicated part of the corresponding strand (see p. 125 for Okazaki fragments in T4).

Lambda DNA can also replicate as part of the bacterial chromosome. To take this optional course, the injected DNA must first direct the synthesis of gene products that promote insertion of the phage chromosome into that of the host at a characteristic site. Also, further genes must be expressed to suppress quite promptly autonomous DNA replication and most other phage functions. When all these conditions are satisfied, the phage DNA is inserted by a process of recombination, entering the state of prophage. All bacterial cells carrying the λ prophage are lysogens and potential centres for the production of λ phage. The genes cI^+, cII^+ and $cIII^+$ are necessary for the efficient maintenance of prophage, though

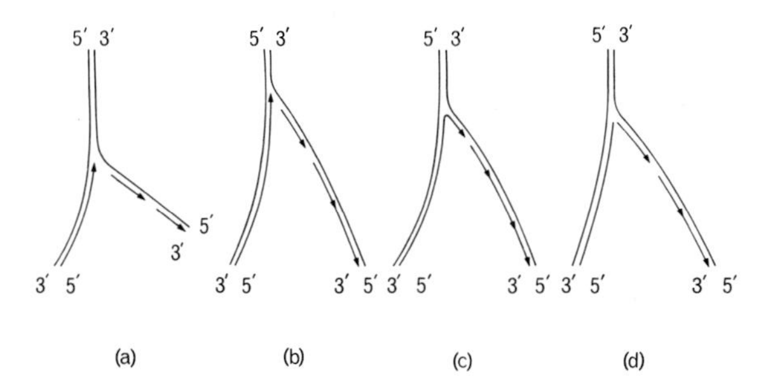

Fig. 7.6 Lambda bacteriophage, illustrating two possible modes of replication of its DNA by the action of DNA polymerase acting only in the 5′ to 3′ direction. In (a) and (b) segments are initiated on the right-hand strand and joined to previously made segments. In (c) and (d) the 5′ to 3′ strand propagates from the left-hand strand past the fork into the right-hand strandand is then nicked at the fork.

the first alone will suffice (genetic map, Fig. 7.7). The gene cI^+ specifies a repressor that prevents expression of other prophage genes and it also represses genes of superinfective λ phage, so rendering the cell immune.

If the repressor ceases to act, the prophage enters on productive growth. This process starts by excision from the bacterial chromosome by a process of recombination the reverse of that by which it entered. Figure 7.8 (a–d) illustrates the mechanism, first proposed by Campbell (1962). There is an attachment site at a specific locus in the lambda chromosome and an homologous locus in the bacterial chromosome. Sometimes excision is abnormal, at the rate of about 10^{-5}, and the resulting phage particles are abnormal in incorporating a short piece of the bacterial chromosome and lacking a piece of the lambda chromosome. Figure 7.8 (d–f) illustrates this for the generation of λ*gal*, which is able to transduce the fragment of the bacterial chromosome into other *E. coli* chromosomes and generate a duplication of the *gal* region. The deficient transducing phage can replicate only if it possesses the termini and is aided by auxiliary λ phages to provide the missing functions. The break points that generate the abnormal phages are unique for each isolate, which is different from any other. The breaks and joins occur between non-homologous regions of phage and bacterial chromosome.

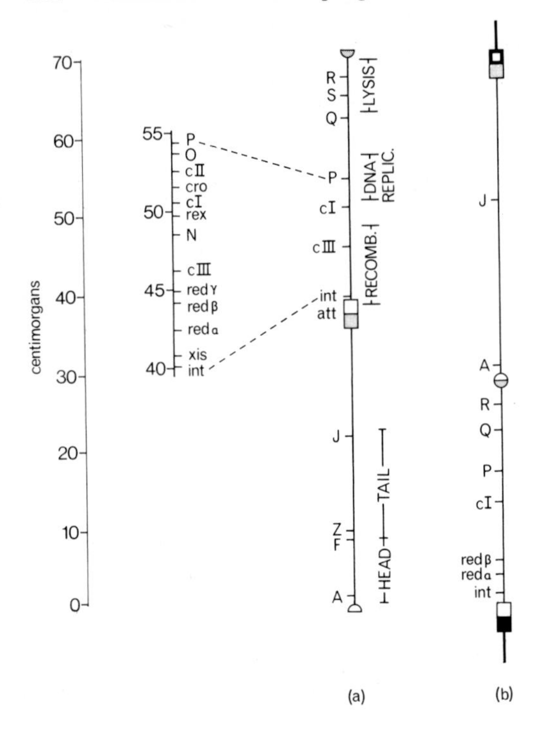

Fig. 7.7 Lambda bacteriophage, linkage maps (a) of lytic phage (b) of prophage integrated into the *E. coli* chromosome, with scale in centimorgans to the left. The loci and the functions, lost in mutants, are shown in the following table:

Locus	Normal gene action
int	integration to make lysogen; also excision.
att	attachment for integration.
xis	complementary to *int* for excision.
redα	exonuclease for general recombination.
redβ	*β* protein for general recombination.
redγ	acts with *redα* and *redβ*.
N	promotes transcription of genes from *int* to *Q*.
Q	promotes transcription of head, tail and lysis genes.
rex	inhibition of growth of T4*r*II mutants.
cI	repressor maintaining prophage, acts at *v2* (near *N*) and at *v1 v3* (right of *cI*)
cII and *cIII*	immunity to lytic phage λ.
cro	decreases action of *cI*, *N*, *redα* and *xis*.
O and *P*	replication of phage DNA.
S and *R*	dissolution of bacterial cell; *R* determines endolysin.
A to *F*	formation of head and its constituents.
Z to *J*	formation of tail.

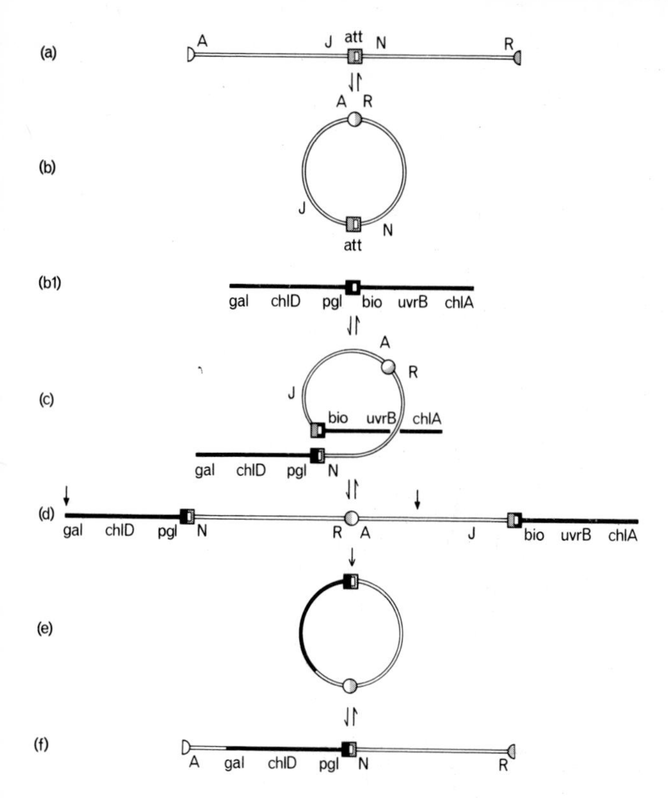

Fig. 7.8 Lambda bacteriophage: (a) to (d) integration into *E. coli* chromosome by recombination between attachment sites, reversal regenerating a free lambda phage chromosome; (d) to (f) genesis of *λgal* transducing phage by non-homologous recombination at the sites marked by arrows. Key to bacterial symbols as Fig. 6.1 and to lambda symbols as Fig. 7.7.

Molecular recombination

Experiments with λ phage (Meselson and Weigle, 1961) indicated for the first time that recombination can result in both strands of part of one chromosome being exchanged for the corresponding part of the other. Two lambda strains, one with heavy isotopes (C^{13}, N^{15}) in its DNA and with normal genes cI^+ and R^+, the other with normal isotopes (C^{12}, N^{14}) and mutant genes *cI* and *R* (referred to as *c* and *mi* originally) were infected simultaneously at a high multiplicity of infection into bacteria with normal isotopes and grown in normal medium. The progeny phages were separated into groups according to the amount of inherited heavy isotopes and also classified for their genotypes. Only a minority contained isotopes from the heavy parent.

Besides the majority of parental normal phages of normal density, there were some with a half heavy chromosome, following semiconservative replication, and a few with a complete heavily labelled DNA molecule. The last are parental DNA molecules that have been repackaged. There were also some types of intermediate density.

Most progeny of the recombinant type $+R$ have light DNA. However, there are also two other sorts of this recombinant, one being nearly heavy (about 90% heavy isotopes) and the other nearly half heavy (about 45% of heavy isotopes). The mean density of the nearly heavy class would be explained if recombination broke both strands of a heavy and a normal molecule at a place randomly located between *cI* and *R* and rejoined the larger heavy fragment to the smaller lighter fragment to give a complete nearly heavy recombinant (Fig. 7.9). The analysis is not fine enough to show whether the breaks in the two strands were or were not at the same site. Breaks at somewhat different sites would allow the formation of a joint molecule preceding covalent bonding. The mean density of the nearly half heavy class is that expected from one round of replication of the nearly heavy class or from a break and rejoin recombination of a half heavy and a normal molecule.

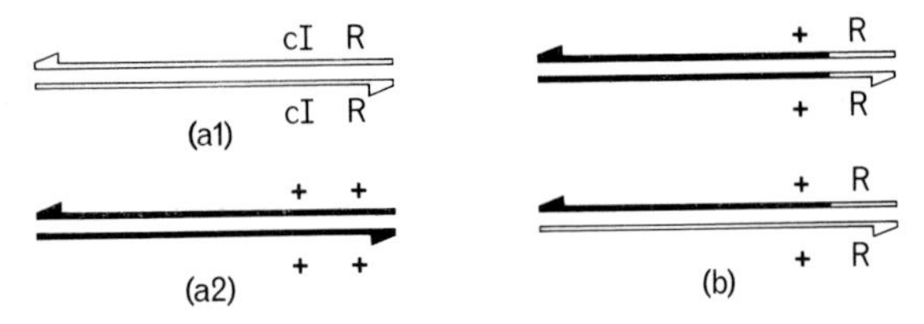

Fig. 7.9 (a) The parents of the cross of λ bacteriophages, analysed by Meselson and Weigle (1961), involving isotopic and genetic differences, the a2 parent having heavy isotopes; (b) certain recombinants with one mutant gene (R) and some heavy isotopes. Gene symbols as in Fig. 7.7.

The recombinant type $cI+$, which has inherited only the right end marker of the heavy parent has a density distribution indicating that only trivial amounts of isotopic label are transferred into it.

Crosses with both parents heavy have used recombination in the interval J to cI in the central third of the map. Fully heavy recombinants are found among the progeny, showing that two heavy parental molecules can break and rejoin.

Transfer of parts of molecules from parent to recombinant offspring is also shown by experiments using lambda phage mutants (λ *b2* and imm^{21}) of different density (Kellenberger *et al.*, 1961). The recombinants λ *b2* imm^{21} and λ $b2^{+}$ are readily separable from one another and from the parents by density gradient centrifugation. When one parent is labelled radioactively, the recombinant progeny have a higher specific radioactivity than progeny of the unlabelled parental genotype. Radioactivity in the latter types of progeny presumably results from recombination outside the map interval *λ b2* to *imm*.

These experiments, and similar ones using host modification as a parental molecular label (Ihler and Meselson, 1963), show that at least some lambda recombination occurs by breakage and rejoining of parental DNA. Moreover, this exchange can occur by each of the rec^+, red^+ and int^+ pathways discussed below. It does not prove that the breaks in either molecule are at precisely the same place, nor that the breaks in the two chains of one molecule are at the same place precisely. Nothing can be concluded about other mechanisms, especially any that might be associated with replicating DNA.

Genetic control of lambda recombination

Three distinct systems affect the recombination of lambda phage genes. At least three genes in *E. coli*, namely $recA^+$, $recB^+$ and $recC^+$, are concerned in recombination in the bacterium (Signer, 1971). Mutation to inactivity in any one leads to mutants which cannot recombine by conjugation, general transduction or episomy (p. 113). At first it was considered that lambda phage showed recombination at approximately normal frequencies in *rec* mutants of *E. coli*, though more careful measurements showed apparent reductions. The lack of serious failure of recombination due to *rec* mutation in the host implied that lambda has a gene or genes that will substitute for the rec^+ genes of *E. coli*. In fact, lambda has two systems red^+ and int^+, specified by genes in lambda itself.

An account of the *E. coli rec* genes is given elsewhere (pp. 115–118). Here it may be noted that *recA* mutants obliterate recombination in *E. coli*, while *recB* and *recC* mutants greatly reduce it. Several other genes are involved including probably *lex* and the suppressors *sbcA* and *sbcB* of *recB* and *recC*. The *recB* and *recC* mutants lack an exonuclease which is specific for double stranded DNA and also an endonuclease specific to single stranded DNA. The *sbcA* suppressor restores recombination to lambda as well as to *E. coli recB* or *recC* mutants.

The red^+ system of lambda comprises two genes, $red\alpha^+$ and $red\beta^+$, and probably a third one, $red\gamma^+$. The gene $red\alpha^+$ specifies an exonuclease specific for 5′ termini, while $red\beta^+$ specifies a protein whose function is unknown except that it appears to complex with the exonuclease and, in consequence, increases its affinity for DNA. Phage *red* mutants do not complement *E. coli rec* mutants.

The int^+ system comprises two genes, int^+ and xis^+; int^+ is necessary for integration of lambda into the *E. coli* chromosome, while xis^+ (as well as int^+) is necessary for excision.

Table 7.2 compares the rec^+, red^+ and int^+ systems with respect to their contributions to recombination in two adjacent segments of the lambda chromosome. The *J cI* segment includes the site (*att*) by which the phage

Table 7.2 The action of the rec^+, red^+ and int^+ systems in recombination in lambda bacteriophage; + = normal wild type; 0 = mutant in one or more genes in the system. The data (Signer, 1971) cover two segments of the lambda chromosome (see Fig. 7.8).

Systems			Recombination %	
Phage		Host		
red	*int*	*rec*	*J–cI*	*cI–R*
+	+	+	7.5	3.6
+	0	0	4.1	3.0
0	+	0	2.0	0
0	0	+	1.3	1.3
+	+	0	7.8	3.1
0	0	0	0	0

attaches to the bacterial chromosome, while *cI R* does not. Recombination in the *J cI* segment is dependent on all three systems, but in the *cI R* segment only the rec^+ and red^+ systems are concerned. Integration of the lambda phage into the host is dependent on the int^+ system and occurs even if the rec^+ and red^+ systems are defective. Conversely, mutants defective in int^+ show normal general recombination provided that the red^+ system is active. Moreover, general recombination due to the red^+ system is almost as great in a defective *rec* host as in a normal one. However, the rec^+ system of the host is only about a tenth as effective if the red^+ system is defective.

There is evidence that recombination due to the int^+ system is reciprocal, whereas that due to the red^+ system is generally not reciprocal. Weil (1969) examined lambda progeny from single bacterial cells, arising from a mixed infection of a *rec* host with two strains of lambda differing at the *susA*, *b2* and *cIII* loci. In the *susA b2* segment only the red^+ system operates, whereas in the *b2 cIII* segment about three quarters of the recombinants are due to the int^+ system. Reciprocal recombinants in the *susA b2* segment were hardly correlated, whereas in the *b2 cIII* segment there was correlation; the correlation coefficients were, respectively, 0.16 and 0.64.

8

Theories of the mechanism of recombination

'*Eureka.*' Archimedes.
'*It is a good rule not to put overmuch confidence in a theory until it has been confirmed by observation. . . . it is also a good rule not to put overmuch confidence in the observational results that are put forward* until they have been confirmed by theory.' Sir Arthur Eddington, 1935, *New Pathways in Science* (Messenger Lectures, 1934).

Classical recombination between factors that are not allelic, i.e. in physiologically distinguishable genes, is typically reciprocal, occurring equally between two strands. However, recombination between alleles, i.e. with mutations at different sites of the same normal gene, is usually not reciprocal. This is most clearly shown in Ascomycetes in which all products of each meiosis can be recovered and separately analysed in an ordered fashion. As first shown by Mitchell (1955), aberrant asci showing gene conversions reveal that for short segments of parental chromosomes, recombination is not reciprocal. Parental allelic genes, entering in equal proportion, are recovered in some tetrads in 3:1 rather than 2:2 ratios, or, where octads are the rule, in 6:2 or 5:3 ratios rather than 4:4. This gene conversion is frequently, probably generally in a half of the cases, associated with reciprocal recombination of nearby flanking genes. The evidence is consistent with the view that all reciprocal recombination (crossing over) is associated with events that would lead to conversion if these were detectable. Where conversion and crossing over events occur together, they usually involve two of the four chromatids, more rarely three. Conversion within a locus is polarized, decreasing in frequency from one end to the other, occurring with frequencies that differ widely from one locus to another. The frequency at a given locus is subject to genetic control that appears to be exercised at a particular locus from which, or near which, the event precursory to conversion starts and spreads.

Conversion is properly defined only when all products from each meiosis can be recovered and analysed. However, the characteristics are definable and can be recognized when, as in *Drosophila melanogaster*, only two of the four products are recovered together. Conversion can also be recognized as the cause of a number of characteristics found when random or selected samples of populations from crosses are analysed. Particularly, there are map distortions, 'high negative interference' and non-additivity of map intervals which are found in fine genetic analysis in micro-organisms. They are all manifestations that a particular class of recombinants occurs more or less often than is expected, on some basis, within a selected population. The principle will therefore be adopted of trying to find a unitary theory, by assuming that if conversion and crossing over as observed in Ascomycetes is explained then other instances may be explicable as special cases.

Recombination is a process that occurs between two molecules of DNA at any one place. Each chromosome or chromatid is one molecule of DNA consisting of two chains of opposite polarity. Each reproduces semiconservatively. In eucaryotes, chromosome replication precedes meiosis, i.e. before chromosome pairing and the stages at which physical recombination appears to occur cytologically. However, a small amount of replication is delayed until zygotene and some replication (or repair) of DNA occurs, especially during pachytene, in addition to the premeiotic and zygotene events. There may be some compensatory destruction.

Modern theories of recombination (reviews and symposia: Grell, 1974; Hotchkiss, 1971, 1974; Kushev, 1971; Radding, 1973; Sobell, 1973) assume an interaction between chains of non-sister chromatids, each a DNA molecule, such that chains of each of two DNA molecules form hybrid molecules. Some breakage (dissociation of covalent bonds in chains) is invoked, together with some dissociation or melting or denaturation (separation apart) of complementary chains, new association between non-sister chains, with the precision of association assured by the sequence of complementary nucleotides. Sometimes there is some erosion by exonucleases, especially following endonucleolytic action to eliminate mismatched bases, with new synthesis to repair gaps making use of the information provided by intact chains and, finally, the closure of nicks between adjacent nucleotides in the chains. The participation of enzymes having the diverse properties of endonuclease, 'denaturase', exonuclease, polymerase and ligase is required, though some theories do not make use of the whole repertoire. Diverse enzymic properties may be associated with one and the same protein which may be specified by several genes.

Since it is evident that enzymes must be involved in catalysing the various events and since it is axiomatic that enzymic action is specific to particular substrates, consideration must be given to the specificities exhibited by the substrate DNA. Each chain of the DNA consists of a series of nucleotides, in each of which a base (adenine, thymine, guanine or cytosine, as a rule) is attached to the 1′ position in deoxyribose, while a phosphoryl group is attached to the 3′ position in one direction and the 5′ position in the other. The main chain is a linear, unbranched polynucleotide with a recurring 3′, 5′ internucleotide linkage joining many thousands or millions of nucleotides. The two chains of a DNA molecule are twisted around one another and held together by hydrogen bonds between pairs of bases, adenine with thymine and guanine with cytosine. The two chains are opposite in polarity, their 3′ to 5′ directions running conversely. Thus at any one place the two chains are not equivalent with respect to a particular enzyme. As examples: (1) an endonuclease may cut the link between guanine and cytosine in one chain but not between cytosine and guanine in the sister chain; (2) exonuclease II operates in the 5′ to 3′ direction attacking the 5′ end and liberating 5′ mononucleotides but not the reverse; (3) polymerase I operates only in the 5′ to 3′ direction adding nucleotides to 3′ hydroxyl groups.

It may be noted that to secure an exchange between two DNA molecules at least four cuts and four joins, or their equivalents, are needed ordinarily. Moreover, unless the cuts are at corresponding places in the strands the exchanges will be unequal with the result that gaps and redundancies would appear in the molecules requiring, respectively, the action of polymerase and exonuclease to restore equivalence. These points are illustrated diagrammatically in Fig. 8.1 for the case where strands of like polarity are cut and joined. In this diagram and others in this chapter, the situation is greatly simplified by omitting the three-dimensional character of the associations and the spiral relation of

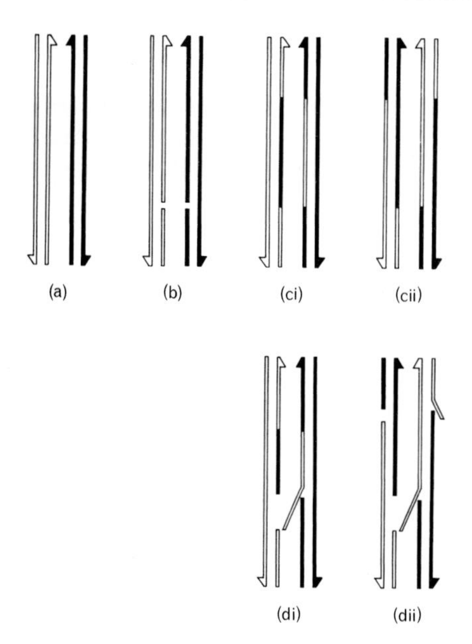

Fig. 8.1 Diagram to show how segments of hybrid (heteroduplex) DNA (c) may or may not be flanked by recombinant neighbouring segments and (d) how non-corresponding break positions lead to gaps and redundancies that may be repaired by polymerase or exonuclease.

complementary sister strands. The barb on each strand indicates the 5′ end of the DNA chain.

It may be noted that even if cuts are initially in strands of unlike polarity, reunions must be between segments of like polarity. Further, it is possible to devise systems involving breaks at different places without actual gaps or redundancies ever existing, because the events of removal and reconstruction go on *pari passu* with breakage and reassociation. The variables covered in different theories are essentially whether cuts occur in strands of the same polarity or of opposite polarity, whether the positions of cuts are matched or not matched in the two chains which interact initially and whether only one chain, rather than two, is cut initially, an extreme form of unmatched cuts but with somewhat different consequences. Significant reviews of data, theories and aspects of their consequences, not specifically discussed in the following treatment, include Gutz, 1971a; Hastings, 1975; Stadler, 1973; and Whitehouse and Hastings, 1965.

8.1 Two copolar strands active initially

The archetype of theories of this class was the proposal of Holliday (1962, 1964). He suggested that cuts occur in paired strands at defined points in chains of like polarity in opposite homologous chromatids (Fig. 8.2). The places at which cuts occur were originally supposed to be links between adjacent DNA molecules but, more likely, they are special sequences of nucleotide pairs which may be recognized by a specific endonuclease. After breakage, the broken strands unravel from the intact ones on one side of the breaks, pair with non-sister intact strands and then reattach in a new way to the free ends of the links. The two chromatids are now joined by a half chiasma (Fig. 8.2c). The actual cross link is a three-dimensional structure in which the parental strands play an equal part. The process is completed by the precise breakage and reunion of non-complementary strands at the point where the strands exchange partners. Thus there is no deletion nor duplication produced in any strand.

If the strands which take part in the resolution are not the two in which the original breaks occurred, the half chiasma is converted to a whole chromatid chiasma, recombining regions flanking the event. Otherwise, if resolution is in the two strands in which the original breaks occurred, no chiasma is formed and there is no recombination of flanking regions. Presumably, there is an equal chance for the alternatives. In both cases there is a segment of hybrid DNA in each of the chromatids. If the hybrid segment spans a site of difference between the parents, mispairing of bases will occur, a different mispair in each of the segments. It is postulated that this condition of mispaired bases is unstable. By some means one of the mismatched bases at each site is removed and replaced by a base which pairs correctly with the remaining one. In this way a wide range of possible conversions may occur (see page 156 *et seq.*). Kushev (1971) has suggested that correction of a heteroduplex is biased toward the recipient strand, the donor strand being identified as that with a free end.

The polarized occurrence of conversion is explained by Holliday as a consequence of the initiation of hybridity of DNA from a fixed site and interference in its spread by the first site of heterozygosity. However, this does not accord with the prevalence of co-conversion which suggests that the establishment of an extensive segment of hybrid DNA is not interfered with by sites of heterozygosity. Nor does it account for the fact that reciprocal recombination between two allelic sites of difference is virtually always accompanied by recombination of flanking regions.

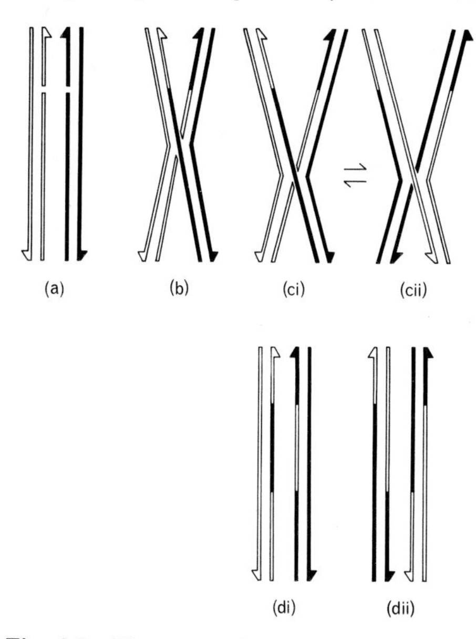

Fig. 8.2 Diagram to illustrate Holliday's (1964) hypothesis. Breaks occur at corresponding sites in non-sister strands of like polarity (a). Following denaturation and exchange of partner (b) a half chiasma is formed and this may migrate (c). Breakage at the half chiasma in either of the isomeric forms of (c) leads to completion of recombination (d), with unlike segments of heteroduplex in each chromatid and with crossing over in dii.

Sigal and Alberts (1972) constructed a molecular model of the Holliday cross stranded structure, the half chiasma. This exercise showed that it could be built with satisfactory bond lengths and angles and with no bases unpaired. The half chiasma can migrate either way by rotation of both duplexes in the same sense. Migration of the half chiasma extends the region of heteroduplex DNA symmetrically on both chromatids. The driving force for movement is unknown. Meselson (1972) has shown that rotary diffusion, arising from rotation of both duplexes in the same sense about their helical axes, will drive the half chiasma far enough and quickly enough to generate quite long segments of hybrid DNA in a relatively short time.

A mechanism proposed for transformation in bacteria is essentially similar. Fox (1966) suggested that a single strand from the end of an incoming piece of DNA from a donor found an unpaired region in the DNA of the recipient and there paired with the appropriate strand (Fig. 8.3). Depending upon where the jointed molecule is cut, it could produce a heterozygote either of the parental or of the recombinant type with respect to the neighbouring parts of the recipient chromosome. The DNA that is displaced or is in excess from the donor may be degraded either before or after removal from the intermediate structures. Only the parental type (Fig. 8.3a–di) seems appropriate for transformation, but the processes leading to recombination around the heteroduplex are also appropriate to phage and other recombination.

The mechanism suggested for transformation could be adapted to explain recombination in λ phage, if a single stranded end were produced by nicking and unwinding, followed by its pairing with one strand of an unpaired region in another phage chromosome. Cutting of various

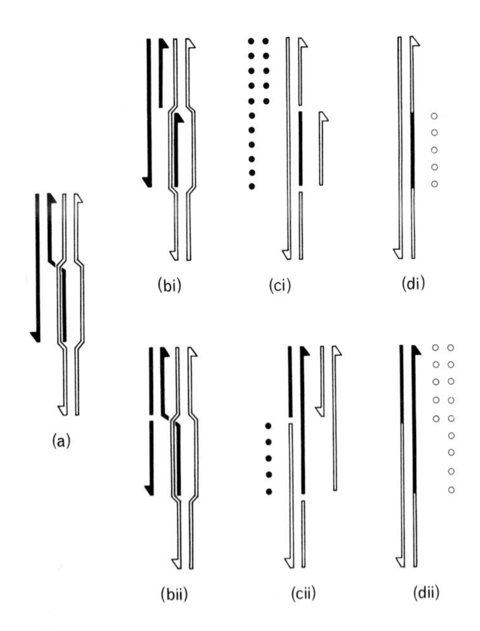

Fig. 8.3 Diagram (a to di) of hypothesis suggested by Fox (1966) to account for transformation in bacteria. One strand of an invading fragment pairs with the complementary strand of the host's chromosome in a region where the DNA chains are denatured. Breaks are induced and ligation occurs. The mechanism could also lead to crossing over (bii to dii).

strands can then generate a heteroduplex flanked either by parental DNA or by recombinant DNA (Fig. 8.4). Formally, this mechanism differs from the Holliday theory mainly in having the initial nick in one strand of only one parental homologue, rather than in both. The consequence is a half cross strand (a quarter chiasma) rather than a full cross strand. From the point of view of λ phage, a way to avoid the recombination fouling two circles would be required.

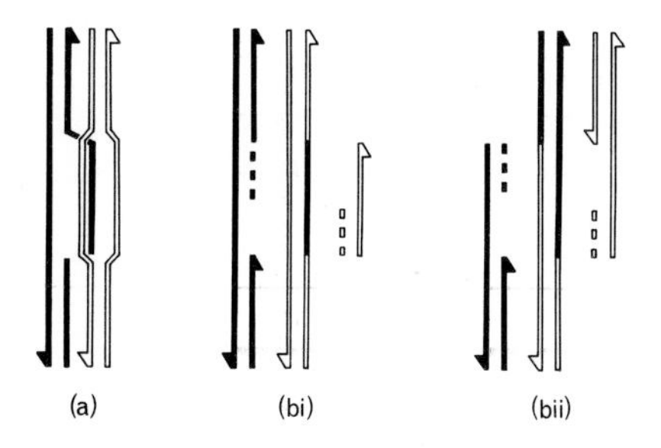

Fig. 8.4 Application of Fox's hypothesis to recombination between two phages. Some synthesis in the 5′ to 3′ direction is indicated by broken lines.

Cross and Lieb (1967) suggested, for the special case of insertion or excision of λ phage, into or out of the *E. coli* chromosome, that homologous regions come into close proximity and that there is an exchange of hydrogen bonding between complementary bases over a fairly short region. The result would be the formation of two complementary half chiasmata betwen which there has been exchange of partner strands. The structure is identical to the configuration achieved by Sobell (see Fig. 8.10f) by a different route. The nexus is resolved by cuts in non-complementary pairs of chains immediately either side of the exchanges, such that pairs of opposite polarity are cut. The nicks are closed by ligase, but it is envisaged that exonuclease might enlarge the gap and so require polymerase action.

Fogel and Hurst (1967) suggested that, following pairing of homologous chromatids, there may be breakage of one or both non-complementary strands at a site of heterozygosity. It is assumed that the base differences between the interacting chromatids create stress. Each broken strand now anneals with the complementary strand of the homologue giving rise to a half chiasma and to mispaired bases in each chromatid. Resolution proceeds by either of the paths suggested by Holliday after 'terminalization' to the end of the gene locus, which is assumed to be a discontinuity.

There being no apparent end to possible variations, it is appropriate to consider here a proposal by Stahl (1969) designed to allow gene conversion in the molecular joint other than by correction of mismatched base pairs (Fig. 8.5). Recombination is initiated by local DNA synthesis to form 'sex circles', before any breakage occurs. The sex circles are considered to be formed at fixed loci. Next there is a break and join exchange of non-sister arcs of the sex circles, followed by a second exchange between non-sister arcs. Finally excess arcs or portions of them are removed to yield the completed recombinant structures. The mechanism allows two non-reciprocal exchanges to form products that are reciprocally recombinant for regions outside the circles, provided that the second exchange obeys an *ad hoc* rule of 'good sense'. This is that the material is saved that allows the construction of two complete progeny molecules. There is no obvious molecular basis. Without the rule recombination can be non-reciprocal. Two of the possible pairs of exchange are shown in Fig. 8.5c.

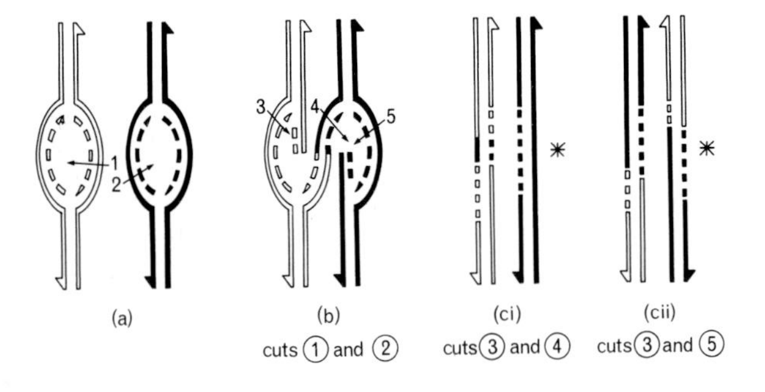

Fig. 8.5 Diagram of the sex circle hypothesis of Stahl (1969) which allows gene conversion other than by formation of heteroduplex and correction of mismatched bases. Homologous regions in two chromatids are denatured and new synthesis occurs in them. There follows a series of breaks and regions which conserve two complete chromatids in which one segment (*) is derived from one parental chromatid, appearing as conversion.

Holliday and Whitehouse (1970) have analysed this hypothesis in the light of available data. They show that the following do not agree with the predictions of Stahl's hypothesis: the patterns of polarity in relation to outside markers, inequality of direction of conversion at a site, the frequencies of post-meiotic segregation and map expansion.

8.2 Two antipolar strands active initially

Whereas Holliday postulated that the primary breaks occur in strands of the same polarity, Whitehouse (1963) assumed that the initial cuts occur in strands of opposite polarity, i.e. in complementary strands one from each of two homologous chromatids. Two alternatives were discussed originally, one starting with breaks at nearly but not exactly homologous points (Fig. 8.6), and the other with initial breaks at exactly corresponding positions (Fig. 8.7).

With exchange initiated by nicks that are not exactly opposite each other, the parts of two broken chains are assumed to uncoil from their sisters in opposite directions (Fig. 8.6a) and to associate and coil round one another so that complementary base pairing between their terminal parts results in the formation of a joint molecule (Fig. 8.6b). New synthesis occurs from the other broken ends alongside the old intact chains and these in turn uncoil from their complementary parents and unite by complementary base pairing to form another joint molecule (Fig. 8.6c). The two joint molecules cross one another to form a structure similar to a chiasma, but with nicks in various places. The old nucleotide chains which have remained intact and of which two complementary replicas have been made break down (Fig. 8.6d); the cross strands unite by nucleotides filling the gaps between the two joint molecules and the proximal and distal parts of the chain in which a gap has been eroded (Fig. 8.6e).

The synthesis leading to the second joint molecule could begin before the broken strands have unwound and paired. This synthesis could even

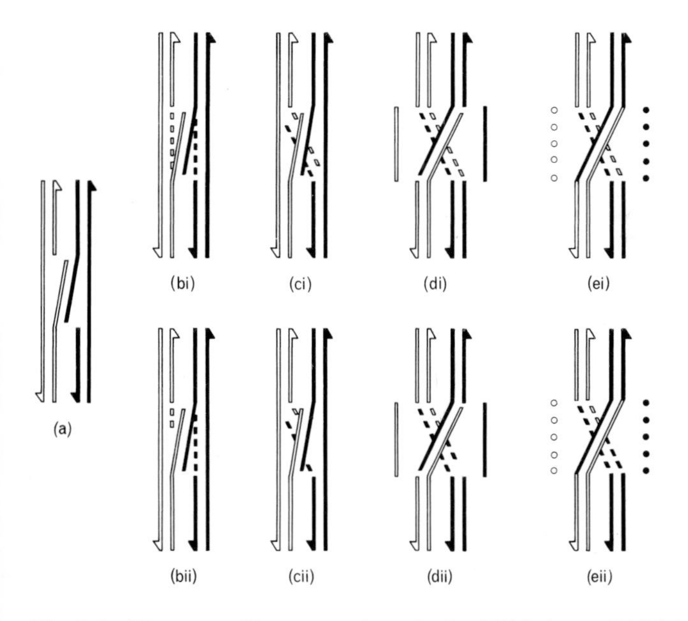

Fig. 8.6 Diagram to illustrate an hypothesis of Whitehouse (1963) in which the initial breaks, in non-sister chains of unlike polarity, are at non-homologous sites. Denaturation, followed by new pairing (b) and resynthesis of these segments, equally (bi) or unequally (bii), followed by their pairing, leads to an incipient chiasma (d) and a complete one (e). The heteroduplexes in the two recombinant chromosomes are alike.

drive the unwinding by displacing the strands that will eventually form the joint molecule. Resynthesis on one exposed parental strand that is incomplete at the time of hybridization (Fig. 8.6cii) could be continued subsequently (Fig. 8.6dii) by copying the extension from the newly synthesized strand of opposite parentage. This could occur for both newly synthesized strands. Both pathways, Fig. 8.6bi–ei and bii–eii, give recombinants that are reciprocal for outside markers. In both there is a segment of hybrid DNA which need not correspond exactly in the two recombinant molecules and may vary in length from one cross over to another. The recombinants from Fig. 8.6bii to eii can be non-reciprocal for any genetic sites in the lately synthesized region. This theory does not account for conversion without a flanking cross over.

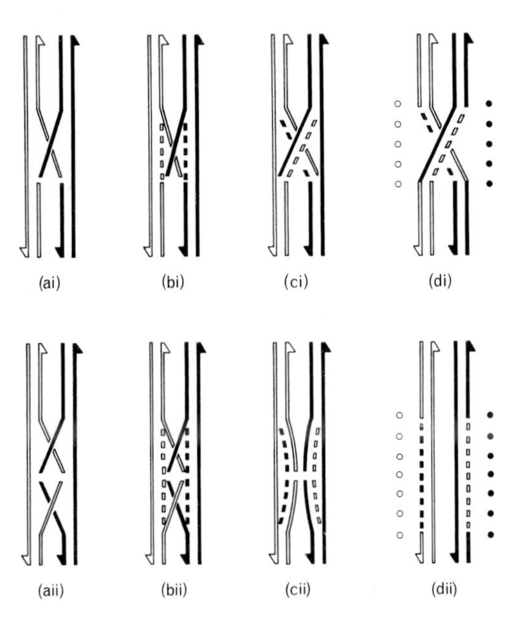

Fig. 8.7 Diagram to illustrate an hypothesis of Whitehouse (1963) in which the initial breaks are at homologous sites in non-sister chains of unlike polarity. Depending upon whether denaturation occurs on one side or both sides of the break points, the formation of heteroduplexes in each recombinant chromosome may be accompanied by a cross over (di) or not (dii).

The alternative theory does so (Fig. 8.7). In it the nicks were initially proposed to be exactly opposite and later (Whitehouse and Hastings, 1965) to be at fixed points in homologous positions, though this is not strictly necessary. The difference from the previous theory is that, after the initial breaks and strand separation (Fig. 8.7a), chain synthesis of a discrete segment occurs alongside the unbroken chains (Fig. 8.7b) and joint molecules arise by base pairing between complementary old and new broken strands (Fig. 8.7c). Breakdown of parental strands, from which copies were taken and which were previously unbroken, together with ligation of nicks, would complete the process (Fig. 8.7d). If the broken chains separate only on one and the same side of the initial nicks, the segments of hybrid DNA would be flanked by a cross over (Fig. 8.7ai–di). However, if the broken chains separated from their intact complementary sisters on both sides of the nicks and then paired with the newly made complementary fragment strands, there would be hybrid DNA in both recombinant molecules, but no crossing over of flanking regions (Fig. 8.7aii–dii). It may be noted that the hybrid DNA is alike in the two chromosomes.

Hastings (1973) has suggested that the primary breaks for meiotic recombination are produced by those segments which failed to replicate in the last pre meiotic S-phase, but did so at zygotene. At a given locus, each homologue would have a gap in each of its chromatids, but in DNA chains of opposite polarity; these may be referred to as α and β chromatids. On a Holliday mechanism, a half chiasma could be formed between the alphas (or betas) of two homologues. On a Whitehouse mechanism, exchange would occur between an α of one homologue and a β of the other. In both cases, only two pairs of chromatids, mutually exclusive, may form cross overs. Since the DNA chain giving rise to an α or a β is continuous over long distances, only two strand or four strand double cross overs could arise by either mechanism alone. However, the observed ratio (1:2:1) of 2-strand to 3-strand to 4-strand could be accounted for by an equal mixture of the two types of mechanism.

Several variants have been proposed, particularly in relation to events in particular procaryotes. Boon and Zinder's (1971) theory begins, like Whitehouse's first one (Fig. 8.6), with breaks at different levels in two non-sister strands of opposite polarity (Fig. 8.8) and the formation of a

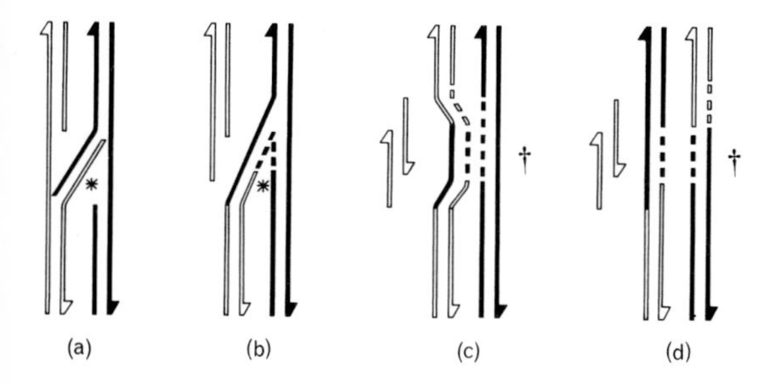

Fig. 8.8 Diagram to illustrate the hypothesis of Boon and Zinder (1971) suggested for *f*1 bacteriophage. Initial breaks are at non-homologous sites in non-sister chains of unlike polarity. Pairing results in the formation of a replicating fork (*) which leads to a segment of one chromatid being present in both recombinant chromosomes (†).

single cross joint together with a replication fork by which a section of one chromosome is duplicated. One of the copies will replace the corresponding part of the chromosome. This results in gene conversion. At the other end of the cross joint a break in the remaining intact stand is followed by closure to the cross joint (Fig. 8.8b). Termination involves joining one of the two arms of the replicating fork to the other chromosome. This may involve a 'simple return' (Fig. 8.8c), in Boon and Zinder's terminology, or a 'cross return' (Fig. 8.8d). One yields a conversion in one chromosome, spanned by non-recombinant parental material, the other a conversion in one chromosome accompanied by a reciprocal recombination of flanking regions. There is also a heteroduplex segment in both cases.

Broker and Lehman (1971) supposed that complementary broken strands on the same side of the nicks pair, move outward and expose gaps (Fig. 8.9a). As the joint moves outward, the intact pair of strands move together and pair (Fig. 8.9b). It is then possible to move branches along each other without loss of base pairing. The movement is continued until additional nicks or gaps are reached, at least three (Fig. 8.9c) being needed to resolve the moving branch system (Fig. 8.9d). Figure 8.9d illustrates the result of a combination of nicks (in Fig. 8.9c) which leads to a cross over spanning the hybrid region. Nicks in the other two strands at the upper end of Fig. 8.9c would result in a non-cross over arrangement spanning the hybrid section. In an alternative form of the theory an exonuclease removes the branch strands not involved in the hybrid region pairing, allowing annealing of the exposed complementary strands and moving the branch points. This continues until nicks, randomly present in the strands, are reached.

Taylor (1967) bases a hypothesis on a scheme of replication of chromosomes occurring in discrete replicons. Each replicon is assumed to

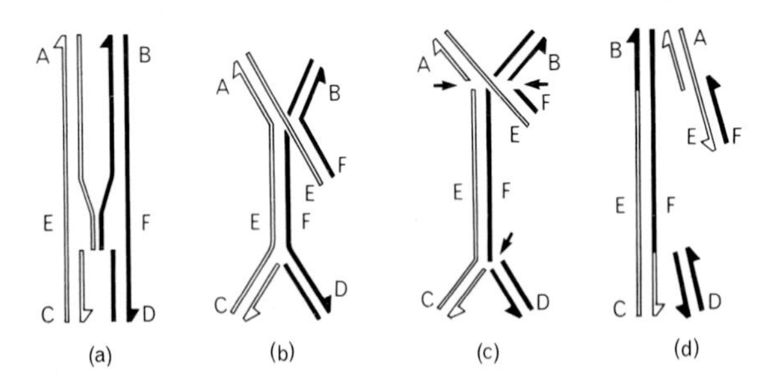

Fig. 8.9 Diagram to illustrate method of formation of branched molecules in bacteriophage T4, according to hypothesis of Broker and Lehman (1971). Breaks occur at homologous sites in non-sister chains of opposite polarity. Denaturation, followed by new pairing, occurs and the fork is able to migrate (b). It will do so until further nicks are encountered (c) that will allow further molecular changes. The one illustrated gives a cross over flanking a heteroduplex (d).

be bounded by a 'replicating guide', not an actual discontinuity in the DNA molecule but somehow attached to it and holding bits together. Replication occurs independently in each replicon and proceeds in the 5′ to 3′ direction beginning first at one end of the replicon and a little later starting at the other end. Each of the original chains is also assumed to be broken near its 5′ end within each replicon. This is applied to recombination by assuming pairing between strands of opposite polarity, derived from homologues. The system will generate heteroduplexes but all are bounded by cross overs.

Sobell (1972) proposed that there are specific regions (promoters) placed between every locus or operon along the chromosome. These promoters are self complementary because they have symmetrically arranged base sequences on each chain and are able to form hair pin associations (Gierer-like) that may have open loops at the ends due to regions that are not self complementary (Fig. 8.10c). A fundamental feature of the theory is the formation of a synaptic structure (Fig. 8.10f)

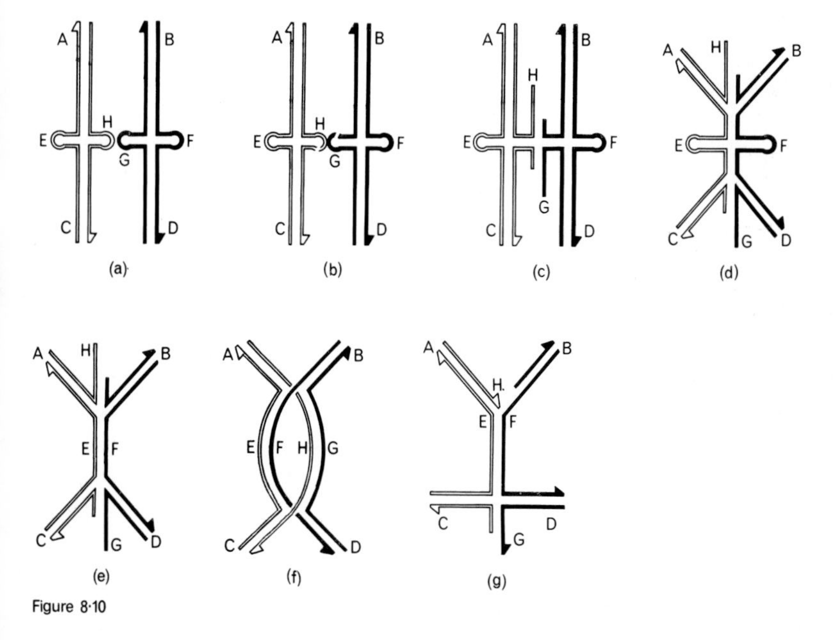

Fig. 8.10 Diagrams to illustrate the genesis of a synaptic complex (f) by the mechanism proposed by Sobell (1972). Following local denaturation in regions with self complementary sequences within DNA chains, these pair (a) leaving terminal loops. Breaks occur in two of these (G and H) at non-homologous sites (b) and pairing occurs between complementary chains (c). Migration of branch points (d and e) may lead to a Broker–Lehman structure (g) or, with pairing between H and G and ligation, to the synaptic complex (f) which can generate recombinants (Fig. 8.11).

at these places, capable of migration into structural genes and capable of resolution by a series of endo- and exonucleolytic steps to effect recombination. Conversion of the promoters from the double helix state to the Gierer structure is assumed to be caused by a specific protein (or proteins). The central sequences that do not possess symmetry are exposed to the action of endonuclease, either randomly or with specificity. When complementary loops are nicked, the homologous Gierer structures can associate by base pairing (Fig. 8.10c) followed by propagation of the hybrid DNA to form the intermediate shown in Fig. 8.10d. In the absence of ligase action, this can be converted to the Broker–Lehman branched structure (Fig. 8.10g) by migration. With ligase activity, a Sobell synaptic structure having two Holliday half chiasmata, is produced (Fig. 8.10f). It results from base pairing of homologous single strands (G, H) followed by sealing of nicks by ligase. It has twofold symmetry. Hence it can be recognized by an endonuclease having twofold symmetry that could simultaneously cut strands of the same polarity at homologous sites. Depending upon which pairs of strands are cut at each half chiasma, the same or different, there will either be no exchange or a reciprocal exchange respectively, of flanking markers. The hybrid structure also has the property of migrating in either direction along parental DNA molecules. In this way the extension of hybrid DNA and the opportunity of recombination can spread to any part of the genome.

The migratory heteroduplex will explain polarity in conversion of single sites and co-conversion (Fig. 8.11). Taking, for example, the data

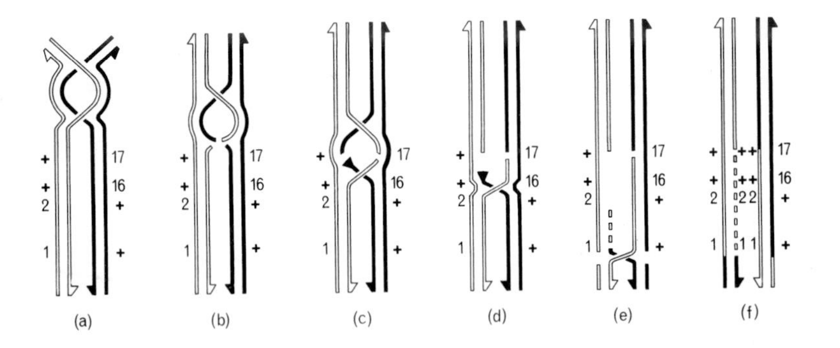

Fig. 8.11 Diagrams to illustrate polarity in conversion and co-conversion starting from Sobell's synaptic complex (Fig. 8.10). The diagram takes as text conversion at the *arg4* locus of yeast in a heterozygote different at four sites in the locus (a). The migratory synaptic complex encounters the first site (17) of difference and triggers nicking of two strands of like polarity by an endonuclease (b) which also acts as an exonuclease degrading one strand (c and d). This continues until homologous strands (of the other polarity in the case shown) are nicked and repair begins (e) and is completed by ligation (f).

of Hurst, Fogel and Mortimer (1972), given in Table 2.10, a migratory heteroduplex proceeding from the distal end of the *arg4* locus will first encounter the site of difference of mutant 17 (Fig. 8.11b). This is assumed to trigger attack by the endonuclease causing nicks in homologous strands. It is also assumed that this nuclease also has exonucleolytic activity, degrading one of the two nicked strands. Subsequent migration proximally of the migratory duplex causes the distal half of it to fall apart, so fixing the starting point at which hybrid DNA begins in the final recombinant molecules (Fig. 8.11c, d). Still further migration, terminated by nicking of homologous strands, results in the formation of a heteroduplex of variable length in one chromosome, a homoduplex formed by polymerase repair in the other (Fig. 8.11e), the whole being spanned by parental segments which may (Fig. 8.11f) or may not be recombined. The heteroduplexes, of different length in different cells, are assumed to be repaired following excision of one strand over the whole heteroduplex region.

8.3 Both strands of each DNA molecule active initially

To explain recombination in various bacteriophages, it has been proposed that the first step is the breakage of both strands of each molecule (or chromosome) but at different places in different homologues (Figs 8.12, 8.13). Certain restriction endonucleases, such as *H. influenzae* II and *H. aegypticus*, act on both strands at the same point because of the reverse reciprocal character of the two chains around the point in question. Meselson and Weigle (1961) suggested for λ phage that broken molecules with homologous segments pair by association of complementary strands in the overlap after these have dissociated from one another (Fig. 8.12b). Recombinant molecules are completed by DNA synthesis that is structurally like two normal replication forks proceeding in opposite directions. This hypothesis does not account for the formation of a heterodup-

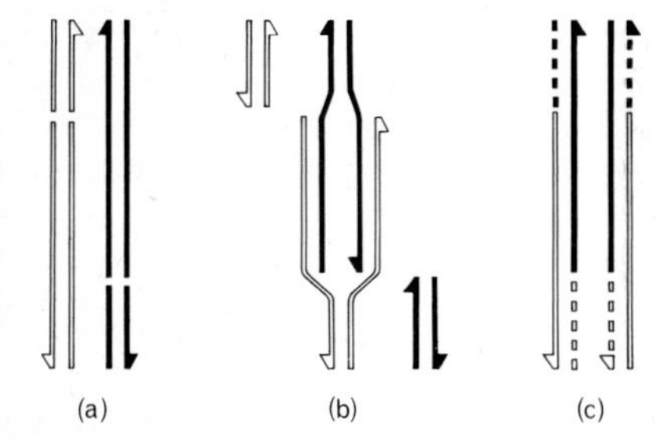

Fig. 8.12 Diagram to illustrate method of recombination by breakage of both strands of two homologues, proposed by Meselson and Weigle (1961) for lambda phage. Following breakage (a) sister strands denature and non-sisters pair (b). Replication at the forks leads to recombinants each with a heteroduplex segment.

lex flanked on both sides by parental material from the same molecule. This can be achieved only by a second exchange. The hypothesis also does not account for the observation that nearly heavy recombinants are formed from heavy parents (see page 134) crossed to light parents and grown in light medium, unless such a double event is invoked.

If the molecules are linear initially the products would each be equivalent to a parental molecule in size. But if the molecules were circular before breakage, the products would be linear and duplicated for all except the overlap. A return to a circle could come about by pruning back under the influence of exonucleases until complementary parts of sister strands at each end of the molecule were exposed.

Thomas (1966, 1967) proposed for T2 and T4 phages that digestion of the broken ends by exonuclease exposes single strand regions which can pair (Fig. 8.13). Such associations may leave gaps that can be filled in by

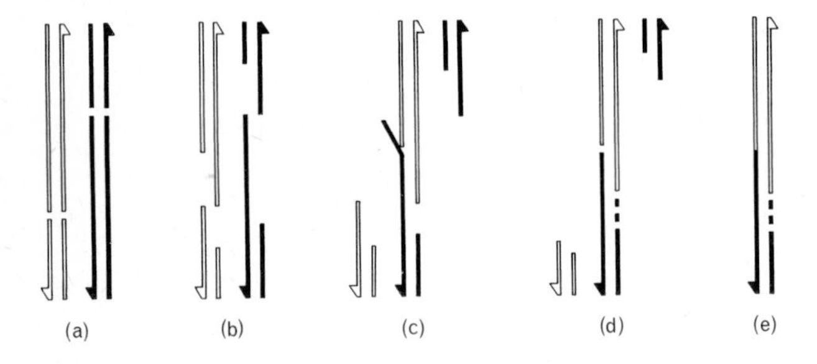

Fig. 8.13 Diagram to illustrate method of recombination by breakage of both strands (a) of two homologues, proposed by Thomas (1967) for T2 and T4 phages. Digestion by exonuclease in the 5′ to 3′ direction exposes single strand regions (b) that can pair (c). Repair, by polymerase and exonuclease action (d) and ligase (e), leads to recombinant molecules that contain a heteroduplex.

synthesis and also excess pieces that can be digested away. Finally the gaps between adjacent nucleotides, equivalent to nicks in strands, would be sealed by a ligase. Only one recombinant product is produced and it necessarily has a heteroduplex flanked by DNA from two different parental molecules. The enzymic requirements are relatively simple, but the process seems extravagant in the wastage of DNA and it is difficult to produce recombinants having a heteroduplex segment in an otherwise parental molecule, except by an unlikely combination of two exchanges.

8.4 Only one strand active initially

The equal formation of heteroduplex DNA on both chromatids should be reflected in the genetic data. The occurrence of aberrant 4:4 segregations in *Sordaria* (Kitani *et al.*, 1962) shows that heteroduplex DNA can form at the same site on both chromatids. Moreover, at the *b2* locus in *Ascobolus*,

Leblon and Rossignol (1973) found that the ratios of 6:2, 5:3 and aberrant 4:4 octad were in good agreement with expectation on the assumption that heteroduplex DNA usually forms on both chromatids.

Nevertheless, the observed results of recombination require the assumption of asymmetrical processes of resolution, unless it is presupposed that the intermediate structures were themselves asymmetrical. In *Neurospora crass* the *cog*$^{+}$ gene, in the heterozygous state, imposes a preference for conversion in the chromosome in which it lies (see pp. 78–80). Asymmetry is also indicated by observations in *Ascobolus immersus* and yeast (see pp.41–46).

Several hypotheses beginning with a single strand broken in one helix and invading another helix have been proposed. An hypothesis that would allow the formation of both asymmetrical and symmetrical heteroduplex DNA and would account for other properties would be particularly attractive. Sobell's hypothesis generates asymmetry only.

Paszewski (1970) proposed that the invading strand (Fig. 8.14a) was partly degraded, then paired with the complementary strand of the homologue and was extended by synthesis (Fig. 8.14b). In so doing it

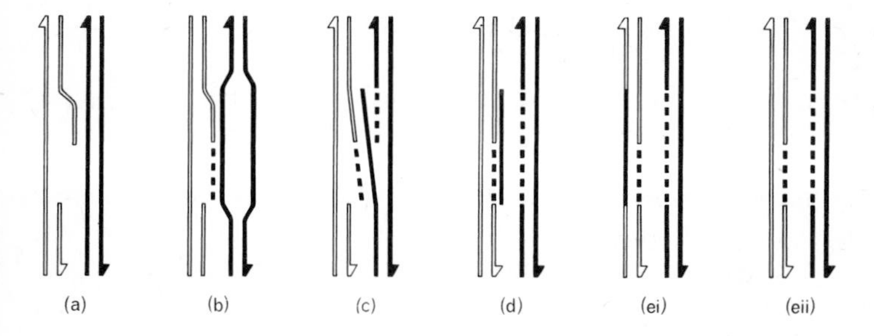

Fig. 8.14 Diagram to illustrate method of recombination involving an initial nick in only one strand that is partly degraded (a) and invades a homologue (b) leading to a joint bridge (c). A heteroduplex in one product (eii) or neither (ei) may result. Hypothesis of Paszewski (1970).

captured a segment of the complementary strand to form a Whitehouse joint bridge (Fig. 8.14c). The original invading strand then turned back to rejoin the original chromosome (Fig. 8.14d), the captured segment of the other chromosome being discarded (Fig. 8.14eii) or incorporated (Fig. 8.14ei). Its place in the homologue was meantime reconstituted by synthesis. The proposals made by Paszewski cover only conversion without crossing over. The hypothesis could no doubt be extended to cover recombination of flanking regions. According to Holliday and Whitehouse (1970) this hypothesis does not account for inequalities in direction of conversion at a site, nor for map expansion.

Hotchkiss (1974), on the other hand, proposed that strand displacement occurred by 3′-ward extension by polymerase (Fig. 8.15a, b), likely to build sufficient distortion to stimulate eventual nicking by repair enzymes of the strand of like polarity (Fig. 8.15c). This would liberate

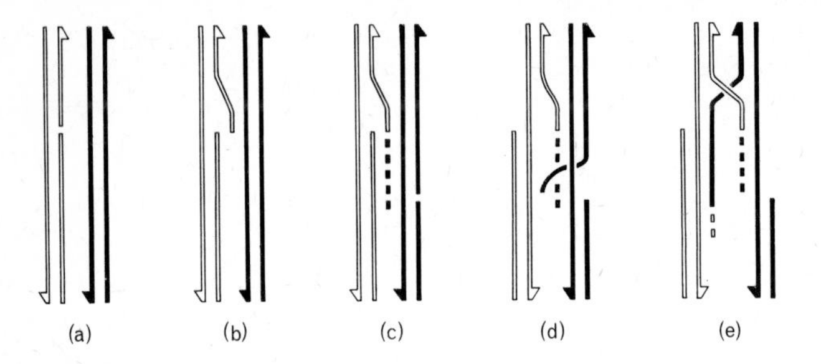

Fig. 8.15 Diagram to illustrate the hypothesis of Hotchkiss (1971) which invokes a nick in only one strand (a) the 3′ end of which then invades the homologue (b) and extends by replication using the information from the complementary strand (c). This induces a nick in the other strand of the homologue (c) and this, in turn, invades the first homologue (d) and so a half chiasma is established.

another 3′ end and trigger an equivalent return attack by it further along the original helix also in the 3′ direction (Fig. 8.15d), producing a Holliday half chiasma (Fig. 8.15e). This would be resolved by further breakage and reunion. These two hypotheses are similar, but differ in the polarity relationships of the strand with the primary and secondary nicks.

Meselson and Radding (1975) have proposed an attractively simple hypothesis which is an adaptation of the Holliday hypothesis. A single break in one strand of one homologue (Fig. 8.16a) becomes the site of displacement by a DNA polymerase. The strand on one side of the break grows, displacing the strand on the other side. The displaced strand pairs with the complementary sequence in the homologue (Fig. 8.16b) and induces a single strand break in the displaced (distorted) strand. Two fragment strands are then ligated, leaving two other fragments one of which continues to grow by polymerase activity while the other is eroded by exonuclease action (Fig. 8.16c). These two enzymic activities are conceived as different catalytic properties of the same complex enzyme. This concerted action results in further strand transfer. It produces a tract of heteroduplex DNA on only one of the two interacting molecules and so is designated as *asymmetrical strand transfer*. The necessary rotations of the duplexes are driven quickly by the chemical energy derived from the polymerization reaction. The strand transfer would move in a definite direction due to the polarized direction of the polymerization. Dissociation of the enzyme allows the structure shown in Fig. 8.16di to isomerize to that in Fig. 8.16dii. In the first structure the arms that flank the site of exchange are parental strands, whereas in the second structure the recombinant strands are outside.

The cross connection (a Holliday half chiasma) in Fig. 8.16dii is free to migrate, presumably in either direction, as a consequence of rotary diffusion of the two DNA molecules. Hence, while heteroduplex DNA was previously formed asymmetrically, strand transfer is now symmetrical and heteroduplex DNA may be formed on both molecules (Fig. 8.16eii). This new structure may also isomerise reversibly to yield the

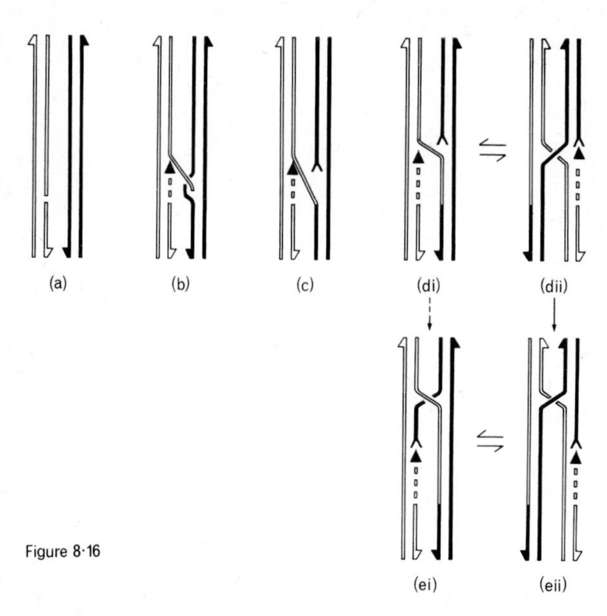

Fig. 8.16 Diagram of hypothesis due to Meselson and Radding (1975). Following a nick in one strand of the primary homologue (a), the 5′ end of the cut strand invades the secondary homologue, inducing a break in (b) and join to the co-polar strand (c). The 3′ end of the first strand nicked grows by polymerase action in the 5′ to 3′ direction, while the 5′ end of the second strand nicked is eroded by exonuclease action (c and d). The structure (di) is capable of isomery to (dii), which has a half chiasma capable of migration (eii) and of isomery to (ei). Resolution of the half chiasma will yield a cross over from (eii) or not from (ei). In both cases there is an extensive heteroduplex in the secondary homologue, with a shorter heteroduplex and a segment of parental type on the primary homologue.

structure in Fig. 8.16ei, which also exchanges the parental and recombinant configurations of the flanking arms. These structures are physically, but not genetically, like the Holliday half chiasmata illustrated in Fig. 8.2.

The hypothesis advanced by Catcheside and Angel (1974) to account for the behaviour of recombination in an interchange heterozygote with one break inside the *his-3* locus is exactly similar to the Meselson and Radding hypothesis. The similarities are that the first break occurs in one strand, some erosion occurring from the 5′ end of the nick and some synthesis from the 3′ side (but not explicit in the original). The 5′ end interacts with the homologous chromosome provoking a break in a strand of like polarity and joining to the 3′ end, leaving the new 5′ end to erode by exonuclease action. This leads to a segment of parental duplex in the chromosome first nicked and a segment of hybrid duplex in the homologue. In the case of conversion at the *his-3* locus a gene cog^+ situated between *his-3* and *ad-3* raises the frequency of conversion (and of crossing over between *his3* and *ad3*) and does so preferentially in the cog^+ strand when cog^+ is heterozygous. In these circumstances the normal rule of polarity is overcome. The effect could be accounted for if the break in the homologue was followed by coupling of its 5′ end to the growing 3′ end of the strand first broken. The hybrid duplex would then be formed by the steps shown in Fig. 8.17, on the chromosome first broken, and would favour conversion in this chromosome.

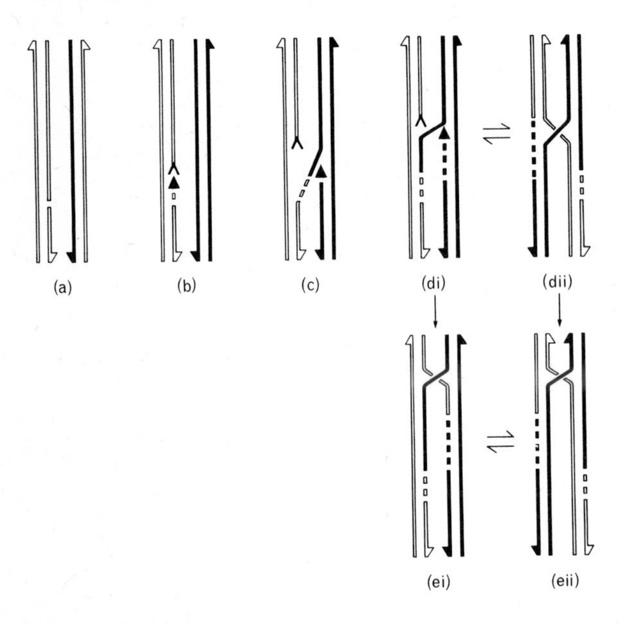

Fig. 8.17 Diagram of a modification, described on p. 156, of the hypothesis due to Catcheside and Angel (1974). It is similar to the hypothesis of Meselson and Radding (Fig. 8.16), but results in the longer heteroduplex (e) being in the primary homologue, first cut (a, b).

8.5 Gene conversion by repair

The most convincing evidence that mismatched base pairs or missing bases in hybrid DNA are substrates for a repair mechanism (or mechanisms) is derived from the observations on gene conversion and post-meiotic segregation in *Sordaria fimicola* and *Ascobolus immersus*. In crosses between a mutant and the normal, the frequency of conversion asci of the two types 6+ :2*m* and 2+ :6*m* is often quite unequal, either type, in different instances, being in the ascendant. This contrasts with the extensive data from yeast that indicate a general parity in direction of conversion, even in the cases where the mutant is a deletion; in *Ascobolus*, correction is strongly biased to filling in a deletion. Again, some mutants show frequent post-meiotic segregation while others do so rarely. Moreover, the frequencies of 5+ :3*m* and 3+ :5*m* are often unlike. Allelic mutants often show these differences in behaviour.

It has often been considered difficult to account for these results on any theory of recombination that assumes gene conversion to involve the erosion and replacement of a length of one chain of hybrid DNA. Nevertheless such an event is necessary to account for the patterns of co-conversion observed in yeast, the interaction of two alleles in *Ascobolus* and for the behaviour of *ade6*-M26 in fission yeast. It is generally assumed that the inequalities are explicable if the particular mismatched pair influences whether or not the repair will occur and also which of the mismatched bases is excised and replaced. In these ways it might appear possible to account for a variety of behaviour of different missense mutants, since eight different kinds of non-complementary base pairs are possible. Thus four different paired combinations of mispaired bases may be encountered. These are: (1) AA+TT; (2) GG+CC; (3) AG+CT; (4) AC+GT. Nevertheless, the experiments of Spatz and Trautner (1970) show that in a particular kind of mispair, judged to be GT, there is no bias to the preferred correction of T to C. The G was corrected to A nearly as often. It therefore seems unlikely that the mispair itself has much influence on the direction of correction.

Analyses of these data have generally assumed symmetrical production of hybrid DNA in two chromatids. However, some experiments show that hybrid DNA is quite often formed in only one chromatid, at least over part of a locus. If there were some genetic difference between parental chromosomes, such as the *cog*$^+$ and *cog* genes in *Neurospora crassa*, that promoted the formation of hybrid DNA more frequently in one chromatid than in its homologue, disparity in 6:2 and 2:6 frequencies could arise from another cause. In Fig. 8.18 the mutant site *a* is assumed to lie in a segment of hybrid DNA formed in one homologue (Fig. 8.18b, c); that the flanking regions are in the parental conformation is irrelevant to the argument. There are also another *a*+ and another +*b* chromatid not drawn. Taking Fig. 8.18b first, correction of *a* to + will give a 6+ :2*a* segregation, correction of + to *a* will give a 4+ :4*a* and no correction until after meiosis will give 5+ :3*a*. The corresponding results for Fig. 8.18c are 4+ :4*a*, 2+ :6*a* and 3+ :5*a*. If the hybrid DNA was formed more frequently in the native chromatid (Fig. 8.18b) than in the homologue and the particular mismatch had no influence on the direction of repair, it would happen nevertheless that 6+ :2*a* would be more common than 2+ :6*a*, while 5+ :3*a* would occur more often than 3+ :5*a*. Causes of inequality from this source could be prevalent in *Ascobolus*, but not in *Sordaria fimicola*. The latter is homothallic and all published data appear to concern the wild strain A1 and mutants derived from it. Nevertheless, asymmetry is probably not universal and part of a locus may show hybrid DNA in one chromatid and another part in both chromatids (Fig. 8.17e). The occurrence of abnormal 4+ :4*m* segregations require symmetry of hybrid DNA at the site concerned.

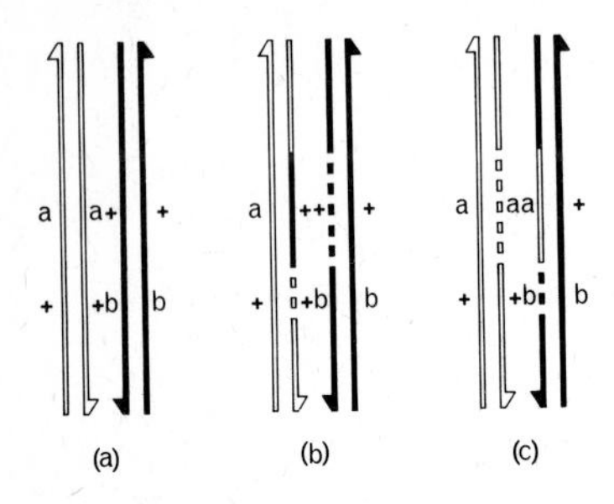

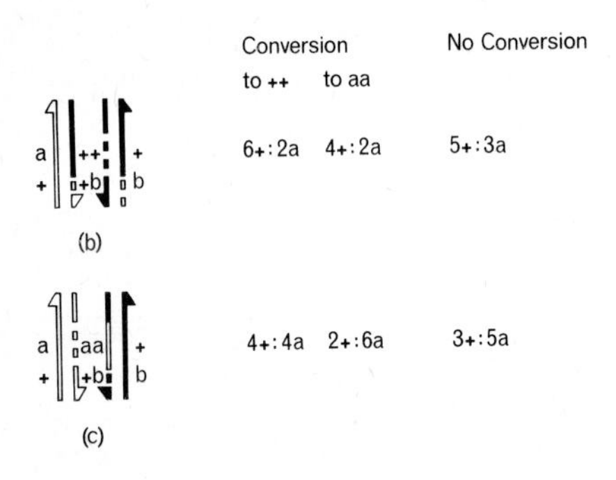

Fig. 8.18 Diagram to show the source of different patterns of conversion in an octad with heteroduplex formed in only one homologue. If a heteroduplex is more likely to be formed in one homologue, due to a *cog* gene difference, inequalities between 6+ :2*a* and 2+ :6*a* (or between 5+ :3*a* and 3+ :5*a*) could result from this influence alone.

At the present time it does not seem possible to predict how a particular mismatch will be corrected. It is evident that the factors involved embrace more than the particular site in question.

Postscript

'Experience is the child of Thought and Thought is the child of Action'.
Disraeli, Earl of Beaconsfield: *Vivian Grey bk. V*, ch. 1.

The present state of knowledge of recombination is similar to that of a jigsaw puzzle of which we possess parts of several copies, each of the same picture, but cut up in different patterns. Moreover, some glimpses of what should be the same part of the picture appear to conflict with one another. The most notable and irritating of these is the contrast between parity in direction of conversion in yeast and bias in direction shown by some other fungi. The problem is whether the nature of the heteroduplex assumed to be an intermediate has any influence on the type of conversion. Deliberate experiments with phage suggest that there is only a limited effect. It is necessary to explore this problem deeply with unselected tetrads to determine whether there are truly departures from parity. If there are, then the rules governing the biases must be found and any genes, and the enzymes they control, must be discovered. At present, the situation is one of uncertainty and it is one about which judgements must be reserved.

The gap that separates eucaryotes from procaryotes is bridged to some extent by the apparent similarities in the intimate molecular events of recombination, at least with respect to the formation of heteroduplexes, the removal of mispaired bases and their neighbours on one DNA chain and consequent repair. It seems likely that enzymes catalysing similar chemical events will be found in all organisms investigated. However, details are likely to be different especially in respect of the conditions under which homologous enzymes act.

In the eucaryotes there is a rigid adherence to the molecular events occurring between four chromatids (the four strands of classical genetics), though apparently the events occur only between two (one of each pair of sisters) in any one short region. This is in contrast to procaryotes where the event appears to occur between two, rather than two of four, DNA molecules. The difference may arise from the eucaryotes possessing more than one chromosome in the haploid set and the consequent need for strict control of sexual processes of syngamy and reduction. Departure from strict control is penalized by sterility, a penalty less severe in procaryotes.

In procaryotes, the systems of mitotic replication and of recombination appear to be mixed together in the sense that they are not essentially separated as they are in eucaryotes, where recombination is nearly restricted to meiosis. Even in eucaryotes the separation is not perfect, for rare recombination occurs between homologues at, or during, mitosis. Entry into meiosis is under genetic control and, like all events of differentiation, involves a switch mechanism. This and the genes and biochemical events directly concerned are poorly understood.

The mechanisms of chromosome pairing at meiosis need to be understood. The problems include those of how close molecular interaction may occur in a synaptinemal complex in which the bulk of the homologous chromosomes are relatively distant from one another and of how the intimate molecular interactions are controlled. In systems of this kind errors could be common and would damage the process by causing sterility. Thus the selection of genetic systems not prone to error would be favoured. Probably there are genetically determined systems for the detection of errors in pairing and for their elimination.

Mechanisms to minimize error must extend to the control of 'illegitimate' recombination, requiring relatively short sequences of homology and, as evidence from bacteria suggests, the action of different systems of genes and the enzymes they determine. Sometimes, as in the numerous systems of variegation encountered in higher plants, there is a release from control, matched now by cases in bacteria. These exceptional kinds of recombination may well throw light on the regular kind.

In conclusion, although it is possible that we understand recombination in a general way, it is certain that much more study of the genetics and enzymology must be undertaken before a definitive and consistent account can be written. Most important in the near future will be thorough analyses of a few suitable organisms, and a careful dissection of the apparent conflicts of evidence.

Bibliography

In the following list of references, the titles of eight periodicals have been abbreviated as follows:

AJBS	Australian J. biol. Sci.
CSH	Cold Spring Harbor Symposia Quantitative Biology
JBC	J. biol. Chem.
JMB	J. molec. Biol.
MGG	Molec. Gen. Genetics
PNAS	Proc. nat. Acad. Sci. U.S.A.
PRS	Proc. Roy. Soc. London series B
ZIAV	Zeits. ind. Absts. u.-Vererb.

AHMAD, A. and LEUPOLD, U. (1973). *MGG*, **123**, 143–58.
ALBERTS, B. M. (1970). *Fed. Proc.*, **29**, 1154–63.
ALBERTS, B. M. and FREY, L. (1970). *Nature*, **227**, 1313–18.
ANDERSON, T. F. (1958). *CSH*, **23**, 47–58.
ANGEL, T., AUSTIN, B. and CATCHESIDE, D. G. (1970). *AJBS*, **23**, 1229–40.
ANRAKU, N., ANRAKU, Y. and LEHMAN, I. R. (1969). *JMB*, **46**, 481–92.
AVERY, O. T., MACLEOD, C. M. and MCCARTY, M. (1944). *J. Exp. Med.*, **79**, 137–58.
BADMAN, R. (1972). *Genet. Res.*, **20**, 213–29.
BALDY, M. W. (1968). *CSH*, **33**, 333–8.
BANDIERA, M., ARMALEO, D. and MORPURGO, G. (1973). *MCG*, **122**, 137–48.
BARANOWSKA, H. (1970). *Genet. Res.*, **16**, 185–206.
BARBOUR, S. D. and CLARK, A. J. (1970). *PNAS*, **65**, 955–61.
BEADLE, G. W. (1932a). *ZIAV*, **62**, 291–304.
BEADLE, G. W. (1932b). *Genetics*, **17**, 481–501.
BEADLE, G. W. (1932c). *ZIAV*, **63**, 195–217.
BEADLE, G. W. (1933). *Cytologia*, **4**, 269–87.
BEATTIE, K. L. and SETLOW, J. K. (1971). *Nature New Biol.*, **231**, 177–9.
BELLING, J. (1928). *Univ. Calif. Publs. Bot.*, **14**, 283–91.
BENZER, S. (1961). *PNAS*, **47**, 403–15.
BERGER, H., WARREN, A. J. and FRY, K. E. (1969). *J. Virology*, **3**, 171–5.
BERNSTEIN, H. (1967). *Genetics*, **56**, 755–69.
BERNSTEIN, H. (1968). *CSH*, **33**, 325–31.
BODMER, W. F. (1965). *JMB*, **14**, 534–57.
BODMER, W. F. (1966). *J. Gen. Physiol.*, **49**, 233–58.
BODMER, W. F. and GANESAN, A. T. (1964). *Genetics*, **50**, 717–28.
BOLE-GOWDA, B. N., PERKINS, D. D. and STRICKLAND, W. N. (1962). *Genetics*, **47**, 1243–52.
BOON, T. and ZINDER, N. D. (1971). *JMB*, **58**, 133–51.
VON BORSTEL, R. C., CAIN, K. T. and STEINBERG, C. M. (1971). *Genetics*, **69**, 17–27.
VON BORSTEL, R. C., QUAH, S.-K., STEINBERG, C. M., FLURY, F. and GOTTLIEB, D. J. C. (1973). *Genetics*, **73**, 141–51.

BRESCH, C., MÜLLER, G. and EGEL, R. (1968). *MGG*, **102**, 301–6.
BRESLER, S. E., KRENEVA, R. A. and KUSHEV, V. V. (1968). *MGG*, **102**, 257–68.
BRESLER, S. E., KRENEVA, R. A. and KUSHEV, V. V. (1971). *MGG*, **113**, 204–13.
BROKER, T. R. and LEHMAN, I. R. (1971). *JMB*, **60**, 131–49.
BROWN, S. W. and ZOHARY, D. (1955). *Genetics*, **40**, 850–73.
CAIRNS, J. (1961). *JMB*, **3**, 756–61.
CAIRNS, J. (1963). *JMB*, **6**, 208–13.
CALLAN, H. G. (1972). *PRS*, **181**, 19–41.
CAMPBELL, A. M. (1960). *Virology*, **11**, 339–48.
CAMPBELL, A. M. (1962). *Adv. Genetics*, **11**, 101–45.
CAMPBELL, A. M. (1969). *Episomes*. Harper and Row, New York.
CARPENTER, A. T. C. (1975). *PNAS*, **72**, 3186–9.
CASE, M. E. and GILES, N. H. (1958a). *PNAS*, **44**, 378–90.
CASE, M. E. and GILES, N. H. (1958b). *CSH*, **23**, 119–35.
CASE, M. E. and GILES, N. H. (1964). *Genetics*, **49**, 529–40.
CASTELLAZZI, M., GEORGE, J. and BUTTIN, G. (1972). *MGG*, **119**, 139–52, 153–74.
CATCHESIDE, D. E. A. (1970). *AJBS*, **23**, 855–65.
CATCHESIDE, D. E. A. (1974). *AJBS*, **27**, 561–73.
CATCHESIDE, D. G. (1944). *Anns. Bot. NS*, **8**, 119–30.
CATCHESIDE, D. G. (1966). *AJBS*, **19**, 1039–43.
CATCHESIDE, D. G. (1974). *Ann. Rev. Genetics*, **8**, 279–300.
CATCHESIDE, D. G. (1975). *AJBS*, **28**, 213–25.
CATCHESIDE, D. G. and ANGEL, T. (1974). *AJBS*, **27**, 219–229.
CATCHESIDE, D. G. and CORCORAN, D. (1973). *AJBS*, **26**, 1337–53.
CATO, A. and GUILD, W. R. (1968). *JMB*, **37**, 157–78.
CHIU, S. M. and HASTINGS, P. J. (1973). *Genetics*, **73**, 29–43.
CHOVNICK, A. (1961). *Genetics*, **46**, 493–507.
CHOVNICK, A. (1973). *Genetics*, **75**, 123–31.
CHOVNICK, A., BALLANTYNE, G. H., BAILLIE, D. L. and HOLM, D. G. (1970). *Genetics*, **66**, 315–29.
CHOVNICK, A., BALLANTYNE, G. H. and HOLM, D. G. (1971). *Genetics*, **69**, 179–209.
CHOVNICK, A., FINNERTY, V. G., SCHALET, A. and DUCK, P. (1969). *Genetics*, **62**, 145–60.
CLARK, A. J. (1971). *Ann. Rev. Microbiol.*, **25**, 438–64.
CLARK, A. J. (1973). *Ann. Rev. Genet.*, **7**, 67–86.
CLARK, A. J. (1974). *Genetics*, **78**, 259–71.
CLARK, A. J., CHAMBERLIN, M., BOYCE, R. P. and HOWARD-FLANDERS, P. (1966). *JMB*, **19**, 442–54.
CLARK, A. J. and MARGULIES, A. D. (1965). *PNAS*, **53**, 451–9.
CLAYBERG, C. D. (1958). *Proc. 10 Int. Congr. Genet.*, **2**, 53.
CLAYBERG, C. D., BUTLER, L., KERR, E. A., RICK, C. M. and ROBINSON, R. W. (1966). *J. Heredity*, **57**, 189–96.
COOPER, A. D., BURGAN, M. W., WHTE, C. W. and HERRMANN, R. L. (1971). *J. Bact.*, **107**, 433–41.
CREIGHTON, H. B. and MCCLINTOCK, B. (1931). *PNAS*, **17**, 492–7.
CROSS, R. A. and LIEB, M. (1967). *Genetics*, **57**, 549–60.
CURTISS, R. (1969). *Ann. Rev. Microbiol.*, **23**, 69–136.
DARLINGTON, A. J. and BODMER, W. F. (1968). *Genetics*, **60**, 681–4.
DARLINGTON, C. D. (1930). *PRS*, **107**, 50–9.
DARLINGTON, C. D. (1934). *ZIAV*, **67**, 96–114.
DARLINGTON, C. D. (1937). *Recent Advances in Cytology*, Ed 2; Churchill, London.
DAVERN, C. I. (1971). *Progress in Nucleic Acid Research and Molecular Biology* (Eds. J. N. Davidson, W. E. Cohn), 229–58. Academic Press, New York.
DELBRÜCK, M. and BAILEY, W. T. (1946). *CSH*, **11**, 33–7.
DOERFLER, W. and HOGNESS, D. S. (1968). *JMB*, **33**, 661–78.
DOERMANN, A. H. (1973). *Ann. Rev. Genetics*, **7**, 325–41.

DOERMANN, A. H. and BOEHNER, L. (1963). *Virology*, **21**, 551–67.
DOVER, G. A. and RILEY, R. (1972a). *Nature New Biol.*, **235**, 61–2.
DOVER, G. A. and RILEY, R. (1972b). *Nature*, **240**, 159–61.
DRISCOLL, C. J. (1972). *Can. J. Genet. Cytol.*, **14**, 39–42.
DRISCOLL, C. J. and DARVEY, N. L. (1970). *Science*, **169**, 290–1.
EBERSOLD, W. T. and LEVINE, R. P. (1959). *ZIAV*, **90**, 74–82.
EDGAR, R. S., DENHARDT, G. H. and EPSTEIN, R. H. (1964). *Genetics*, **49**, 635–48.
EDGAR, R. S. and LIELAUSIS, I. (1964). *Genetics*, **49**, 649–62.
EDGAR, R. S. and LIELAUSIS, I. (1968). *JMB*, **32**, 263–76.
EDGAR, R. S. and WOOD, W. B. (1966). *PNAS*, **55**, 498–505.
EMERSON, S. (1966). *Genetics*, **53**, 475–85.
EMERSON, S. (1969) in *Genetic Organization*, **1**, 267–360 (Ed. E. W. Caspari and A. W. Ravin). Academic Press, New York.
EMERSON, S. and YU-SUN, C. C. C. (1967). *Genetics*, **55**, 39–47.
EPHRUSSI-TAYLOR, H. and GRAY, T. C. (1966). *J. Gen. Physiol.*, **49** suppl., 211–31.
EPSTEIN, R. H., BOLLE, A., STEINBERG, C. M., KELLENBERGER, E., BOY DE LA TOUR, E., CHEVALLEY, R., EDGAR, R. S., SUSMAN, M., DENHARDT, G. H. and LIELAUSIS, A. (1963). *CSH*, **28**, 375–92.
ESPOSITO, M. S. and ESPOSITO, R. E. (1974). *Genetics*, **78**, 215–25.
FAREED, G. C. and RICHARDSON, C. C. (1967). *PNAS*, **58**, 665–72.
FELDMAN, M. (1966). *PNAS*, **55**, 1447–53.
FINCHAM, J. R. S. (1951). *J. Genet.*, **50**, 221–9.
FINCHAM, J. R. S. and HOLLIDAY, R. (1970). *MGG*, **109**, 309–22.
FINK, G. R. and STYLES, C. A. (1974). *Genetics*, **77**, 231–44.
FINNERTY, V. G., DUCK, P. and CHOVNICK, A. (1970). *PNAS*, **65**, 939–46.
FLAVELL, R. B. and WALKER, G. W. R. (1973). *Exptl. Cell Res.*, **77**, 15–24.
FOGEL, S. and HURST, D. D. (1967). *Genetics*, **57**, 455–81.
FOGEL, S. and MORTIMER, R. K. (1969). *PNAS*, **62**, 96–103.
FOGEL, S. and MORTIMER, R. K. (1970). *MGG*, **109**, 177–85.
FOGEL, S. and MORTIMER, R. K. (1971). *Ann. Rev. Genet.*, **5**, 219–36.
FOGEL, S. and MORTIMER, R. K. (1974). *Genetics*, **77**, s22.
FOGEL, S. and ROTH, R. (1974). *MGG*, **130**, 189–201.
FORTUIN, J. J. H. (1971). *Mut. Res.*, **11**, 149–62, 265–77; **13**, 131–6, 137–48.
FOX, M. S. (1966). *J. Gen. Physiol.*, **49**, 183–96.
FOX, M. S. and ALLEN, M. K. (1964). *PNAS*, **52**, 412–9.
FREESE, E. (1957). *Genetics*, **42**, 671–84.
GILBERT, W. and DRESSLER, D. (1968). *CSH*, **33**, 473–84.
GILES, N. H., DE SERRES, F. J. and BARBOUR, E. (1957). *Genetics*, **42**, 608–17.
GILLIES, C. B. (1974). *Chromosoma*, **48**, 441–53.
GILMORE, R. A. (1967). *Genetics*, **56**, 641–58.
GIRARD, J. and ROSSIGNOL, J.-L. (1974). *Genetics*, **76**, 221–43.
GOLDMARK, P. and LINN, S. (1972). *JBC*, **247**, 1849–60.
GOTTSCHALK, W. (1968). *Nucleus, Seminar Vol. 1968* (Ed. A. K. Sharma), 345–61.
GOWEN, J. W. (1933). *J. Exptl. Zool.*, **65**, 83–106.
GRELL, R. F. (Ed.), (1974). *Mechanisms in Recombination*. Plenum Press, New York and London.
GUTZ, H. (1971a). *Genet. Res.*, **17**, 45–52.
GUTZ, H. (1971b). *Genetics*, **69**, 317–37.
HARGRAVE, J. and THRELKELD, S. F. H. (1973). *Genet. Res.*, **21**, 194–204.
HARM, W. (1964). *Mut. Res.*, **1**, 344–54.
HARTWELL, L. H., CULOTTI, J. and REID, B. (1970). *PNAS*, **66**, 352–9.
HASTINGS, P. J. (1973). *Genet. Res.*, **20**, 253–6.
HASTINGS, P. J. (1975). *Ann. Rev. Genetics*, **9**, 129–44.
HASUNUMA, K. and ISHIKAWA, T. (1972). *Genetics*, **70**, 371–84.
HAWTHORNE, D. C. (1955). *Genetics*, **40**, 511–9.
HAWTHORNE, D. C. and MORTIMER, R. K. (1960). *Genetics*, **45**, 1085–100.

HAWTHORNE, D. C. and MORTIMER, R. K. (1968). *Genetics*, **60**, 735–42.
HECHT, N. B. and STERN, H. (1971). *Exptl. Cell Res.*, **69**, 1–10.
HENDERSON, S. A. (1970). *Ann. Rev. Genetics*, **4**, 294–324.
HERMAN, R. K. (1965). *J. Bact.*, **90**, 1664–8.
HERSHEY, A. D. (1946). *CSH*, **11**, 67–77.
HERSHEY, A. D. (1971) (Ed.). *The Bacteriophage Lambda*. Cold Spring Harbor Lab. Monograph.
HERSHEY, A. D. and CHASE, M. (1952). *CSH*, **16**, 471–9.
HERSHEY, A. D. and ROTMAN, R. (1949). *Genetics*, **34**, 44–71.
HINTON, C. W. (1970). *Genetics*, **66**, 663–76.
HOLLIDAY, R. (1962). *Genet. Res.*, **3**, 472–86.
HOLLIDAY, R. (1964). *Genet. Res.*, **5**, 282–304.
HOLLIDAY, R. (1967). *Mut. Res.*, **4**, 275–88.
HOLLIDAY, R. and HALLIWELL, R. E. (1968). *Genet. Res.*, **12**, 94–8.
HOLLIDAY, R. and WHITEHOUSE, H. L. K. (1970). *MGG*, **197**, 85–93.
HORII, Z. and CLARK, A. J. (1973). *JMB*, **80**, 327–44.
HOSODA, J. (1967). *Bioch. Biophys. Res. Comm.*, **27**, 294–8.
HOTCHKISS, R. D. (1971). *Adv. Genetics*, **16**, 325–48.
HOTCHKISS, R. D. (1974). *Ann. Rev. Microbiol.*, **28**, 445–68.
HOTCHKISS, R. D. and GABOR, M. (1970). *Ann. Rev. Genetics*, **4**, 193–224.
HOTTA, Y. and STERN, H. (1971a). *Develop. Biol.*, **26**, 87–99.
HOTTA, Y. and STERN, H. (1971b). *JMB*, **55**, 337–44.
HOTTA, Y., ITO, M. and STERN, H. (1966). *PNAS*, **56**, 1184–91.
HOWARD-FLANDERS, P. and THERIOT, L. (1966). *Genetics*, **53**, 1137–50.
HOWELL, S. H. and STERN, H. (1971). *JMB*, **55**, 357–78.
HURST, D. D. and FOGEL, S. (1964). *Genetics*, **50**, 435–58.
HURST, D. D., FOGEL, S. and MORTIMER, R. K. (1972). *PNAS*, **69**, 101–5.
IHLER, G. and MESELSON, M. (1963). *Virology*, **21**, 7–10.
IHLER, G. and RUPP, W. D. (1969). *PNAS*, **63**, 138–43.
JACOB, F. and MONOD, J. (1963) in *Biological Organization* (Ed. R. J. C. Harris). Academic Press, New York.
JANSEN, G. J. O. (1964). *Genetica*, **35**, 127–31.
JANSEN, G. J. O. (1970). *Mut. Res.*, **10**, 21–32, 33–41.
JESSOP, A. and CATCHESIDE, D. G. (1965). *Heredity*, **20**, 237–56.
KAHN, P. L. (1964). *JMB*, **8**, 392–404.
KELLENBERGER, G., ZICHICHI, M. L. and WEIGLE, J. (1961). *PNAS*, **47**, 869–78.
KENT, J. L. and HOTCHKISS, R. D. (1964). *JMB*, **9**, 308–22.
KIMBER, G. (1974). *Genetics*, **78**, 487–92.
KITANI, Y. and OLIVE, L. S. (1967). *Genetics*, **57**, 767–82.
KITANI, Y. and OLIVE, L. S. (1969). *Genetics*, **62**, 23–66.
KITANI, Y., OLIVE, L. S. and EL-ANI, A. S. (1962). *Amer. J. Bot.*, **49**, 697–706.
KNAPP, E. and MÖLLER, E. (1955). *ZIAV*, **87**, 298–310.
KOZINSKI, A. and FELGENHAUER, Z. Z. (1967). *J. Virol.*, **1**, 1193–1202.
KRUSZEWSKA, A. and GAJEWSKI, W. (1967). *Genet. Res.*, **9**, 159–77.
KUSHEV, V. V. (1971). *Mechanisms of Genetic Recombination*. 'Science', Leningrad Branch.
KUSHNER, S. R., NAGAISHI, H. and CLARK, A. J. (1972). *PNAS*, **69**, 1366–70.
KUTTER, E. M. and WIBERG, J. S. (1968). *JMB*, **38**, 395–411.
LACKS, S. (1966). *Genetics*, **53**, 307–35.
LEBLON, G. (1972). *MGG*, **115**, 36–48; **116**, 322–35.
LEBLON, G. and ROSSIGNOL, J.-L. (1973). *MGG*, **122**, 165–82.
LE CLERC, G. (1946). *Science*, **103**, 553–4.
LEDERBERG, J. (1955). *J. Cell. Comp. Physiol.*, **45**, suppl. 2, 75–107.
LEDERBERG, J. (1957). *PNAS*, **43**, 1060–65.
LEHMAN, I. R. (1967). *Ann. Rev. Biochem.*, **36**, 645–68.
LEMONTT, J. F. (1971a). *Genetics*, **68**, 21–33.

LEMONTT, J. F. (1971b). *Mut. Res.*, **13**, 311–7, 319–26.
LEVAN, A. (1939). *Hereditas*, **25**, 9–26.
LEVINTHAL, C. (1954). *Genetics*, **39**, 169–84.
LHOAS, P. (1961). *Nature*, **190**, 744.
LINDEGREN, C. C. (1932). *Bull. Torrey Bot. Club*, **59**, 119–38.
LINDEGREN, C. C. (1953). *J. Genet.*, **51**, 625–37.
LISSOUBA, P., MOUSSEAU, J., RIZET, G. and ROSSIGNOL, J.-L. (1962). *Adv. Genet.*, **11**, 343–80.
MAGUIRE, M. P. (1968). *Genetics*, **60**, 353–62.
MAKAREWICZ, A. (1964). *Acta Soc. Bot. Pol.*, **33**, 1–8.
MANNEY, T. R. (1968). *Genetics*, **60**, 719–33.
MANNEY, T. R. and MORTIMER, R. K. (1964). *Science*, **143**, 581–3.
MATHER, K. (1933). *Amer. Nat.*, **67**, 476–9.
MCCARTY, M., AVERY, O. T. and TAYLOR, H. (1946). *CSH*, **11**, 177–83.
MELLO-SAMPAYO, T. (1971). *Genet. Iber.*, **23**, 1–9.
MESELSON, M. (1964). *JMB*, **9**, 734–45.
MESELSON, M. (1968). In *Replication and Recombination of Genetic Material* (Ed. W. J. Peacock and R. D. Brock), 152–6. Australian Academy of Science, Canberra.
MESELSON, M. (1972). *JMB*, **71**, 795–98.
MESELSON, M. S. and RADDING, C. R. (1975). *PNAS*, **72**, 358–61.
MESELSON, M. S. and STAHL, F. W. (1958). *PNAS*, **44**, 671–82.
MESELSON, M. S. and WEIGLE, J. (1961). *PNAS*, **47**, 857–68.
MITCHELL, M. B. (1955). *PNAS*, **41**, 215–20.
MITCHELL, M. B., PITTENGER, T. H. and MITCHELL, H. K. (1952). *PNAS*, **38**, 569–80.
MOENS, P. B. (1968). *Canad. J. Genet. Cytol.*, **10**, 799–807.
MOENS, P. B. (1969). *Canad. J. Genet. Cytol.*, **11**, 857–69.
MOENS, P. B. (1970). *Proc. Canad. Fed. Biol. Soc.*, **13**, 160.
MORPURGO, G. and VOLTERRA, L. (1968). *Genetics*, **58**, 529–41.
MORTIMER, R. K. and HAWTHORNE, D. C. (1966). *Genetics*, **53**, 165–73.
MOSES, M. J. (1956). *J. Biophys. Biochem. Cytol.*, **2**, 215–8.
MOSES, M. J. (1968). *Ann. Rev. Genetics*, **2**, 363–412.
MOSIG, G. (1968). *Genetics*, **59**, 137–51.
MOSIG, G. (1970a). *Adv. Genetics*, **15**, 1–53.
MOSIG, G. (1970b). *JMB*, **53**, 503–14.
MURRAY, N. E. (1963). *Genetics*, **48**, 1163–83.
MURRAY, N. E. (1968). *Genetics*, **58**, 181–91.
MURRAY, N. E. (1969). *Genetics*, **61**, 67–77.
NAKAI, S. and MORTIMER, R. K. (1969). *MGG*, **103**, 329–38.
NELSON, O. E. (1962). *Genetics*, **47**, 737–42.
NELSON, O. E. (1968). *Genetics*, **60**, 507–24.
NEWCOMBE, K. D. and THRELKELD, S. F. H. (1972). *Genet. Res.*, **19**, 115–9.
NEWMAN, J. and HANAWALT, P. (1968). *CHS*, **33**, 145–50.
NODA, S. (1975). *Heredity*, **34**, 373–80.
OKAZAKI, R., OKAZAKI, T., SAKABE, K., SUGIMOTO, K. and SUGINO, A. (1968a). *PNAS*, **59**, 598–605.
OKAZAKI, R., OKAZAKI, T., SAKABE, K., SUGIMOTO, K. and KAINUMA, R., SUGINO, A. and IWATSUKI, N. (1968b). *CSH*, **33**, 129–43.
OPPENHEIM, A. B. and RILEY, M. (1966). *JMB*, **20**, 331–57.
OPPENHEIM, A. B. and RILEY, M. (1967). *JMB*, **28**, 503–11.
OZEKI, H. and IKEDA, H. (1968). *Ann. Rev. Genetics*, **2**, 245–78.
PALMER, R. G. (1971). *Chromosoma*, **35**, 233–46.
PARKER, J. H. and SHERMAN, F. (1969). *Genetics*, **62**, 9–22.
PASZEWSKI, A. (1970). *Genet. Res.*, **15**, 55–64.
PAUL, A. V. and RILEY, M. (1974). *JMB*, **82**, 35–56.
PAULING, C. and HAMM, L. (1968). *PNAS*, **60**, 1495–1502.
PEACOCK, W. J. (1970). *Genetics*, **65**, 593–617.

PONTECORVO, G. and KAFER, E. (1958). *Adv. Genet.*, **9**, 71–104.
PRITCHARD, R. H. (1955). *Heredity*, **9**, 343–71.
PUTRAMENT, A. (1964). *Genet. Res.*, **5**, 316–27.
PUTRAMENT, A. (1967). *MGG*, **100**, 321–36.
VAN DE PUTTE, P., ZWENK, H. and RÖRSCH, A. (1966). *Mut. Res.*, **3**, 381–92.
RADDING, C. (1973). *Ann. Rev. Genet.*, **7**, 87–111.
REES, H. and JONES, R. N. (1977). *Chromosome Genetics*. Edward Arnold, London.
RHOADES, M. M. (1946). *Rec. Genet. Soc. America*, **15**, 64.
RICHARDSON, C. C. (1969). *Ann. Rev. Biochem.*, **38**, 795–840.
RICHARDSON, C. C., MASAMUNE, Y., LIVE, T. R., JACQUEMIN-SABLON, A., WEISS, B. and FAREED, G. C. (1968). *CSH*, **23**, 151–64.
RILEY, R. (1968). *Proc. 3rd Int. Wheat Genet. Symp.*, 185–95.
RILEY, R. (1974). *Genetics*, **78**, 193–203.
RILEY, R. and BENNETT, M. D. (1971). *Nature*, **230**, 182–5.
RILEY, R. and CHAPMAN, V. (1958). *Nature*, **182**, 713–15.
RODARTE-RAMÓN, U. S. (1972). *Rad. Res.*, **49**, 148–54.
RODARTE-RAMÓN, U. S. and MORTIMER, R. K. (1972). *Rad. Res.*, **49**, 133–47.
ROMAN, H. (1956). *CSH*, **21**, 175–85.
ROMAN, H. (1963). *Methodology in Basic Genetics* (Ed. W. J. Burdette), 209–27. Holden-Day, San Francisco.
ROMAN, H. (1967). *J. Cell. Physiol.*, **70**, suppl. 1, 116–18.
ROMAN, H. and JACOB, F. (1958). *CSH*, **23**, 155—60.
ROPER, J. A. (1952). *Experientia*, **8**, 14–51.
ROPER, J. A. and PRITCHARD, R. H. (1955). *Nature*, **175**, 639.
ROSSIGNOL, J.-L. (1964). Thesis, Fac. Sci. d'Orsay.
ROSSIGNOL, J.-L. (1969). *Genetics*, **63**, 795–805.
ROTH, R. and FOGEL, S. (1971). *MGG*, **112**, 295–305.
ROTH, T. F. and ITO, M. (1967). *J. Cell Biol.*, **35**, 247–55.
RUSSELL, R. L. (1974). *Genetics*, **78**, 967–88.
SAEDLER, H. and HEISS, B. (1973). *MGG*, **122**, 267–77.
SAEDLER, H., REIF, H. J., HU, S. and DAVIDSON, N. (1974). *MGG*, **132**, 265–89.
SALAMINI, F. and LORENZONI, C. (1970). *MGG*, **108**, 225–32.
SANDLER, L. and LINDSLEY, D. L. (1974). *Genetics*, **78**, 289–97.
SANSOME, E. R. (1932). *Cytologia*, **3**, 200–19.
SCHROEDER, A. L. (1970). *MGG*, **197**, 291–304, 305–20.
SEARS, E. R. (1969). *Ann. Rev. Genetics*, **3**, 451–68.
SHARP, P. A., HSU, M.-T., OHTSUBO, E. and DAVIDSON, N. (1972). *JMB*, **71**, 471–97.
SIDDIQUI, O. H. (1962). *Genet. Res.*, **3**, 69–89.
SIGAL, N. and ALBERTS, B. (1972). *JMB*, **71**, 789–93.
SIGNER, E. (1971). In *The Bacteriophage Lambda* (Ed. A. D. Hershey), 139–74. Cold Spring Harbor Lab. Monograph.
SIMCHEN, G. (1967). *Genet. Res.*, **9**, 195–210.
SIMCHEN, G., BALL, N. and NACHSHON, I. (1971). *Heredity*, **26**, 137–40.
SIMCHEN, G. and STAMBERG, J. (1969). *Nature*, **222**, 329–32.
SIMONET, J. M. (1973). *MGG*, **123**, 263–91.
SIMONET, J. M. and ZICKLER, D. L. (1972). *Chromosoma*, **37**, 327–51.
SMITH, B. R. (1965). *Heredity*, **20**, 257–76.
SMITH, B. R. (1966). *Heredity*, **21**, 481–98.
SMITH, D. A. (1975). *Genetics*, **80**, 125–33.
SMITH, P. D., FINNERTY, V. G. and CHOVNICK, A. (1970). *Nature*, **228**, 441–4.
SMYTH, D. R. (1971). *AJBS*, **24**, 97–106.
SMYTH, D. R. and STERN, H. (1973). *Nature New Biol.*, **245**, 94–6.
SNOW, R. R. (1968). *Mut. Res.*, **6**, 409–18.
SNOW, R. and KORSCH, C. T. (1970). *MGG*, **197**, 201–8.
SOBELL, H. M. (1972). *PNAS*, **69**, 2483–7.
SOBELL, H. M. (1973). *Adv. in Genetics*, **17**, 411–90.

SOBELL, H. M. (1975). *PNAS*, **72**, 279–83.
SOOST, R. K. (1951). *Genetics*, **26**, 410–34.
SPATZ, H. C. and TRAUTNER, T. A. (1970). *MGG*, **109**, 84–106.
STADLER, D. R. (1959). *PNAS*, **45**, 1625–9.
STADLER, D. R. (1973). *Ann. Rev. Genetics*, **7**, 113–27.
STADLER, D. R. and TOWE, A. M. (1963). *Genetics*, **48**, 1323–44.
STADLER, D. R. and TOWE, A. M. (1971). *Genetics*, **68**, 401–13.
STADLER, D. R., TOWE, A. M. and ROSSIGNOL, J.-L. (1970). *Genetics*, **66**, 429–47.
STAHL, F. W. (1969). *Genetics*, suppl., **61**, 1–13.
STAHL, F. W., EDGAR, R. S. and STEINBERG, J. (1964). *Genetics*, **50**, 539–52.
STAMBERG, J. (1968). *MGG*, **102**, 221–8.
STAMBERG, J. and KOLTIN, Y. (1971). *MGG*, **113**, 157–65.
STAMBERG, J. and KOLTIN, Y. (1973). *Genet. Res.*, **22**, 101–11.
STERN, C. (1931). *Biol. Zbl.*, **51**, 547–87.
STERN, C. (1936). *Genetics*, **21**, 625.
STERN, H. and HOTTA, Y. (1973). *Ann. Rev. Genet.*, **7**, 37–66.
STERN, H. and HOTTA, Y. (1974). *Genetics*, **78**, 227–35.
STORM, P. K., HOEKSTRA, W. P. M., DE HAAN, P. G. and VERHOEF, C. (1971). *Mut. Res.*, **13**, 9–17.
STREISINGER, G., EDGAR, R. S. and DENHARDT, G. H. (1964). *PNAS*, **51**, 775–9.
STRICKLAND, W. N. (1958). *PRS*, **149**, 82–101.
SUSMAN, M. (1970). *Ann. Rev. Genet.*, **4**, 135–76.
TAYLOR, A. L. and TROTTER, C. D. (1967). *Bact. Revs.*, **31**, 332–53.
TAYLOR, A. L. and TROTTER, C. D. (1972). *Bact. Revs.*, **36**, 504–24.
TAYLOR, J. H. (1965). *J. Cell. Biol.*, **25**, 57–67.
TAYLOR, J. H. (1967). *Molecular Genetics*, Pt. II, 94–135, Academic Press, New York.
TAYLOR, J. H., HAUT, W. F. and TUNG, J. (1962). *PNAS*, **48**, 190–8.
TAYLOR, J. H., WOODS, P. S. and HUGHES, W. L. (1957). *PNAS*, **43**, 122–28.
THOMAS, C. A. (1966). *Progr. Nucl. Acid Res. Mol. Biol.*, **5**, 315–37.
THOMAS, C. A. (1967). *The Neurosciences; A Study Program* (Ed. G. C. Quarton, T. Melnechuk and F. O. Schmitt), 162–82. Rockefeller Univ. Press, New York.
THOMAS, P. L. and CATCHESIDE, D. G. (1969). *Canad. J. Genet. Cytol.*, **11**, 558–66.
TOMIZAWA, J., ANRAKU, N. and IWAMA, Y. (1966). *JMB*, **21**, 247–53.
UNRAU, P. and HOLLIDAY, R. (1970). *Genet. Res.*, **15**, 157–69.
UNRAU, P. and HOLLIDAY, R. (1972). *Genet. Res.*, **19**, 145–55.
VISCONTI, N. and DELBRÜCK, M. (1953). *Genetics*, **38**, 5–33.
WARNER, H. R. and HOBBS, M. D. (1967). *J. Virology*, **33**, 376–84.
WATSON, W. A. F. (1969). *Mut. Res.*, **8**, 91–100.
WATSON, W. A. F. (1972). *Mut. Res.*, **14**, 299–307.
WEIL, J. (1969). *JMB*, **43**, 351–5.
WESTERGAARD, M. and WETTSTEIN, D. VON (1970). *C.R. Trav. Lab. Carlsberg*, **37**, 239–68.
WESTERGAARD, M. and WETTSTEIN, D. VON (1971). *Ann. Rev. Genet.*, **6**, 71–110.
WHITEHOUSE, H. L. K. (1963). *Nature*, **199**, 1034–40.
WHITEHOUSE, H. L. K. (1965). *Sci. Progr.*, **53**, 285–96.
WHITEHOUSE, H. L. K. and HASTINGS, P. J. (1965). *Genet. Res.*, **6**, 27–92.
WIBERG, J. S. (1966). *PNAS*, **55**, 614–21.
WIBERG, J. S. (1967). *JBC*, **242**, 5824–9.
WILDENBERG, J. (1970). *Genetics*, **66**, 291–304.
WILLETTS, N. S. and MOUNT, D. W. (1969). *J. Bact.*, **100**, 923–34.
WINKLER, H. (1932). *Biol. Zbl.*, **52**, 163–89.
WITKIN, E. M. (1969). *Mut. Res.*, **8**, 9–14.
WOMACK, F. C. (1963). *Virology*, **21**, 232–41.
WOOD, W. B. (1974). In *Handbook of Genetics* (Ed. R. C. King). Van Nostrand Reinhold Co., New York.
WOOD, W. B., EDGAR, R. S., KING, J., LIELAUSIS, I. and HENNINGER, M. (1968). *Fed.*

Proc., **27**, 1160–6.
YANOFSKY, C. (1963). *CSH*, **28**, 581–8.
ZICKLER, H. (1934). *Planta (Arch. wiss. Bot.)*, **22**, 573–613.
ZIMMERMAN, F. K. and SCHWAIER, R. (1967). *MGG*, **100**, 63–76.
ZOHARY, D. (1955). *Genetics*, **40**, 874–7.

Index